We Cry For Our Land

Farm Workers in South Africa

Wendy Davies

OXFAM

British Library Cataloguing in Publication Data

Davies, Wendy
We cry for our land : farm workers in South Africa.
1. South Africa. Agricultural industries
I. Title
338.10968

ISBN 0-85598-143-1

Front cover photograph: Farm labourer on an avocado farm, Northern Transvaal (Steve Hilton-Barber/Afrapix)

Published by Oxfam
274 Banbury Road
Oxford
OX2 7DZ

Designed by Opus Design Group, Oxford
Typeset in Palatino and Helvetica
Printed by Tekprint Ltd.
Printed on recycled paper

The Land Act Song

We are children of Africa,
we cry for our land.
Zulu, Xhosa, Sotho
Zulu, Xhosa, Sotho unite.
We are mad over the Land Act,
a terrible law that allows sojourners
to deny us our land,
crying that we the people
should pay to get our land back.
We cry for the children of our fathers
who roam around the world without a home,
even in the land of their forefathers.

CONTENTS

Johannesburg city centre: Women's Day protest by a member of the Federation of Transvaal Women. (Afrapix)

INTRODUCTION

Sipho Rathebe is 70 years old. He has worked on the same farm in Northern Transvaal as a herdsman for the past 52 years. He is suffering from ill health and exhaustion and would like to give up working, but unless he can continue to provide labour, he and his wife will lose their home. His elder son was killed in an accident on the farm five years ago and his younger son, who has moved to the city, does not want to come back and take his father's place as provider of labour on the farm. Sipho earns R30 (about £7.50) a month and a bag of mealies, and is allowed to keep his herd of goats on the farm.

Margaret Mkhize works on a pineapple plantation in Eastern Cape near the border with the Ciskei 'homeland'. Each day she is picked up in the Ciskei and transported in an open truck with the rest of the seasonal workforce, consisting of women and children, to the plantation. Her working day averages ten hours, and another hour is spent travelling. She is allowed a 30-minute break for lunch and two other 15-minute breaks during the day. Her monthly wage is R27.50 (£6.90).

Willem Botha is a wine farm labourer in Western Cape. Aged 45, he looks at least 20 years older. Every working day he receives seven tots of wine at intervals throughout the day. He is also paid in wine for overtime work. He is so dependent on alcohol that he regularly informs on fellow workers in order to get extra wine. His cash wage is R50 (£12.50) a month.

These are not exceptional cases, but entirely typical of the lives of the estimated six million people who make up South Africa's farm workers and their families. Even by the standards of apartheid's oppressive and ailing systems, farm labour in South Africa is particularly exploitative and anachronistic. The Deputy Minister of Population Development, Luwellyn Landers, has described farm workers as being no better off than slaves.[1]

Whereas industrial workers have been able to organise themselves and win concessions from employers over the last two decades, farm workers – with few exceptions – remain entirely at their employers' mercy. They have virtually no legal protection. Poverty wages, long hours, sub-standard accommodation, and exposure to dangerous machinery and chemicals are all commonplace, and workers' problems are compounded by having little or no education, and very little help from outside. This is not to say that conditions are in practice as bad as this for all farm workers. The important point is that the system allows such abuses and that they are very common.

In 1982 the South African government asked the National Manpower Commission to investigate the working conditions of farm workers and domestic workers. The Commission submitted its report in 1984, the government passed it on to the farmers' unions to look at, and since then it has remained under wraps. It is widely believed to be highly critical of farm labour practices and to recommend sweeping changes.

The long delay in publishing and acting on the report reflects the strength of the farmers' lobby within South African politics. Since the 1930s farmers have been favoured by state policies which give them cheap credit, substantial subsidies, and protection from high interest rates, and set quotas on the imports of foreign produce. In return the government has been able to rely on a high proportion of the farmers' votes. However, since the late 1970s the ruling National Party has sought to win more support from the urban business community and has begun to reduce its subsidies to farmers. As the government moves towards reform, an increasing number of farmers are defecting to the right-wing Conservative Party. Today, the quickening pace of change and the prospect of government negotiations with the black majority is sending more and more farmers into the arms of the neo-Nazi Afrikaner Weerstands Beweging (AWB).

There are signs that the government may be prepared to cut its losses with the more conservative elements of the farming community (while retaining the support of the more 'liberal' farmers) and may at last enact some basic legislation to protect farm workers. How far it will be prepared to go remains to be seen.

The lack of published research means that the main focus of this book is on conditions within so-called 'white South Africa'. There is little mention of conditions in the bantustans, or so-called 'homelands', where white agribusiness and 'state'-run agricultural corporations have taken over most of the fertile land, employing workers for low wages and leaving black peasant farming on degraded land at sub-subsistence levels. The book draws on the experience of organisations working with farm workers, including some supported by Oxfam, within the Republic, and on a small number of published sources. It ranges over various agricultural sectors and various aspects of farm workers' lives, drawing attention not only to abuses and exploitative practices but also to battles fought and victories – however limited – won by farm workers and their supporters in their struggle for justice on the farms.

> We condemn the lack of legal protection for farm workers and the gross exploitation which exists on many farms. We demand that farm workers be allowed full trade union rights. We speak directly to the magistrates in the courts in all the rural areas. Today, we are drawing your attention to the terrible miscarriages of justice which have gone on for decades.
>
> *(Nelson Mandela, speaking at a rally of the African National Congress, Pietersburg, 25 March 1990)*

1 DISPOSSESSION AND CONTROL

> (The Khoi Khoi) strongly insisted that we had been appropriating more and more of their land, which had been theirs all these centuries, and on which they had been accustomed to let their cattle graze, etc. They asked if they would be allowed to do such a thing supposing they went to Holland, and they added: It would be of little consequence if you people stayed here at the fort, but you come right into the interior and select the best land for yourselves ...
>
> *(From the diary of Jan van Riebeeck, leader of the expedition which established a Dutch colony in the Cape in 1652)*

The plight of rural blacks in South Africa has roots stretching far back into the past. It begins with the dispossession of black pastoralist and farming communities from the land by successive waves of white settlers from the mid-seventeenth century onwards. Whites gained 'ownership' of new territory in frontier wars with black chieftainships and kingdoms, and by drawing up spurious treaties and deeds of sale whose terms were vastly in their own favour. The settlers' advance was slow and often met with fierce resistance, but in the end superior armed force enabled them to occupy most of the land.

At the same time as appropriating land, British and Dutch ('Boer') settlers needed labour for their farms, plantations and mines. As blacks were unwilling to give up their own independent production, indentured labour was brought from India and China to some areas in the nineteenth century – but not enough to meet the needs of white expansion. Slaves were imported from West Africa until the abolition of slavery by the British in 1834. It was to avoid the new British law that large numbers of Boers headed north in the Great Trek, wresting yet more land from the black inhabitants.

After the abolition of slavery in the Cape, farmers began to complain about the lack of voluntary labour and to seek ways of binding labourers to the farms. Gradually the view hardened that only if blacks toiled for the white man could their presence be tolerated. The British imperialist, Cecil John Rhodes, pronounced a typically inflexible view in 1894: 'It must be brought home to them that in the future, nine-tenths of them will have to spend their lives in daily labour, in physical work, in manual labour.'

As well as forcing blacks to work for white farmers by confiscating their land and destroying their livelihood, the government introduced taxation in order to drive people into the cash economy. Around the turn of the century, however, when white farmers had gained control of virtually all the land, black farmers began to enter into share-cropping or tenancy arrangements with their white 'owners', paying either half their crop or a cash rent in exchange for the right to cultivate a plot of land. In this way large numbers of black peasants managed to remain productive farmers on the land where they had always lived.

Map of South Africa, showing the so-called 'homelands', or bantustans.

Eastern Transvaal, 1989: Former farm worker Christina Nkosi, who was a girl at the time of the Anglo-Boer War, about to be evicted from the farm where she was born. (Lesley Lawson/Afrapix)

The 1913 Land Act

The racial division of land was given the force of law in the 1913 Natives' Land Act, which allocated a mere seven per cent of the land to the black population. (This was extended to 13 per cent in the 1936 Native Land and Trust Act.) Even more devastating for black farmers was the fact that the Act made share-cropping and rent tenancy contracts illegal. Over a million black tenants were forced to leave their homes and tramp the dusty roads with their dying stock in a desperate search for white farmers who might be prepared to break the law and take them on as rent tenants or share-croppers.

The system introduced by the government to replace cash rents and share-cropping was that of labour tenancy, under which people were allowed access to some agricultural land in return for working half the time for the landowner. It reflected the determination of white farmers to destroy the obvious prosperity of many African peasant farmers. The ideology was that blacks should be servants or employees, not farmers.

An eloquent testimony to the devastating effects of the Land Act is given by Sol Plaatje in *Native Life in South Africa*. As a young black journalist, Plaatje travelled through the country, talking to destitute families whose lives had been shattered by the 'demon law', and documenting what he saw and heard. The case of Kgobadi was typical of the suffering of evicted tenants all over the country:

> Kgobadi, for instance, had received a message describing the eviction of his father-in-law in the Transvaal Province, without notice, because he had refused to place his stock, his family, and his person at the disposal of his former landlord, who now refused to let him remain on his farm except on these conditions. The father-in-law asked that Kgobadi should try and secure a place for him in the much dreaded 'Free' State as the Transvaal had suddenly become uninhabitable to natives who cannot become servants; but 'greedy folk hae lang

> airms', and Kgobadi himself was proceeding with his family and his belongings in a wagon, to inform his people-in-law of his own eviction, without notice, in the 'Free' State, for a similar reason to that which sent his father-in-law adrift. The Baas had exacted from him the services of himself, his wife and his oxen, for wages of 30s a month, whereas Kgobadi had been making over £100 a year, besides retaining the services of his wife and of his cattle for himself. When he refused the extortionate terms, the Baas retaliated with a Dutch note, dated the 30th day of June 1913, which ordered him to 'betake himself from the farm of the undersigned, by sunset of the same day, failing which his stock would be seized and impounded, and himself handed over to the authorities for trespassing on the farm'. (....)
>
> Kgobadi's goats had been about to kid when he trekked from his farm; but the kids, which in halcyon times represented the interest on his capital, were now one by one dying as fast as they were born and left by the roadside for the jackals and vultures to feast upon. This visitation was not confined to Kgobadi's stock. Mrs Kgobadi carried a sick baby when the eviction took place, and she had to transfer her darling from the cottage to the jolting ox-wagon in which they left the farm. Two days out the little one began to sink as the result of privation and exposure on the road, and the night before we met them its little soul was released from its earthly bonds. The death of the child added a fresh perplexity to the stricken parents. They had no right or title to the farm lands through which they trekked: they must keep to the public roads – the only places in the country open to the outcasts if they are possessed of a travelling permit. The deceased child had to be buried, but where, when, and how?
>
> This young wandering family decided to dig a grave under cover of the darkness of that night, when no one was looking, and in that crude manner the dead child was interred – and interred amid fear and trembling, as well as the throbs of a torturing anguish, in a stolen grave, lest the proprietor of the spot, or any of his servants, should surprise them in the act. Even criminals dropping straight from the gallows have an undisputed claim to six feet of ground on which to rest their criminal remains, but under the cruel operation of the Natives' Land Act little children, whose only crime is that God did not make them white, are sometimes denied that right in their ancestral home.[2]

The Land Act signalled the end of black independent farming outside the reserves. After the National Party came to power in 1948 and instituted the policy of apartheid, it sought 'historical' justification for the allocation of only 13 per cent of the land to black South Africans. The answer was a falsification of history: a propaganda campaign claiming that blacks had never settled more than this amount of land 'since white and black [had] arrived almost simultaneously' in South Africa. Yet Pretoria knew that black people had settled roughly half of South Africa before the whites arrived. [3]

By the 1950s share-cropping and rent tenancy had been virtually eliminated and replaced by either labour tenancy or wage labour, both of which cost the farmer virtually nothing. Influx controls meant that farm dwellers could not seek alternative employment in towns, and remained bound to the farms as long as the farmer wanted them. With a virtually captive labour force, farmers had excessive control and power, and the scope for exploitation was almost unlimited.

Changes in farming methods in the decades that followed, together with an increasingly sophisticated apparatus of apartheid, would mean worse was to come for the majority of farm workers.

Cornfields, Natal, November 1988: a community threatened with removal from land they have owned since 1910. (Gill de Vlieg/Afrapix)

2 WHITE AGRICULTURE AND BLACK LABOUR

Restructuring and its effects[4]

The last several decades in South Africa – particularly since the Second World War – have seen a transformation in commercial agriculture, which in turn has had major effects on the numbers of farm workers employed and on their working and living conditions. South African agriculture has been restructured by three major means: mechanisation and other technological innovations, the concentration of larger tracts of land in the hands of fewer people, and a high degree of state support and subsidy for these processes.

The technological changes in agricultural production are often equated with mechanisation. In fact, the advent of tractors is merely the most visible of these changes, which have also included such introductions as chemical fertilisers and pesticides. Mechanisation, although still not as widespread as in many other 'developed' countries, has increased considerably since the 1960s. The extent varies widely according to the type of farming and stage in the production process. Maize (which accounts for 45 per cent of cultivated land in South Africa) and wheat are the only two cash crops to have been completely mechanised, while some plantation agriculture remains labour-intensive for most production stages – planting, weeding, cutting, picking – and is only mechanised at the final stages – stacking, loading and packing.

Mechanisation has been accompanied by the concentration of land. The number of farms has steadily decreased (the greatest reduction happening during the 1970s) and the size of farms increased correspondingly. In other words, more land – and more capital – is in the hands of a smaller number of individuals and companies. Between 1950 and 1984, the number of farm owners declined by 40 per cent, to a total of just over 70,000.

The state has been a very important factor in the restructuring process. From the early 1960s it tried both to raise productivity and reduce farmers' dependence on black labour. Concerned about the so-called 'blackening of the platteland' and the exodus of whites from rural areas, it aimed to keep whites in the agricultural sector by improving productivity and profitability. A range of incentives to encourage mechanisation was introduced, including cheap credit and subsidies on fuel and fertiliser. At the same time the state placed restrictions on the number of blacks allowed on any one farm, and took various measures to destroy labour tenancy and replace it by straight wage labour. The 1960s and 1970s were the time of massive forced removals to the bantustans. Farm workers, the majority of whom were labour tenants, were particularly vulnerable to the state's brutal intervention as they possessed no title deeds. Between 1960 and 1982, according to the calculations of the Surplus People Project, over 1.1 million farm workers were expelled from their homes on white farms – a third of all removals of black people during this period.[5]

Forced removal for a community from Marianhill, Natal, 1981. (Omar Badsha/Afrapix)

As the Surplus People Project report points out, 'In many respects, this massive movement of agricultural workers and small peasant producers off the land is not unique to South Africa. It has been a feature of the capitalisation of agriculture world-wide. The intervention of the state to force this movement into the bantustans and away from the urban areas ... is, however, specific to South Africa.'

The state has continued to give substantial support to farmers through subsidies and credit, as well as research and training facilities, extension services, and a quota system to limit imports of foreign produce. However, since the mid-1970s it has turned to the private sector for help in sharing the cost of agricultural support and development, and this has meant large corporations gaining increasing influence and control within the sector's productive core. A number of large corporations, including some multinational companies, have diversified from industry and mining into agriculture. On some corporately owned farms, workers have succeeded in winning certain concessions and improvements in their working conditions – but they are in a minority.

South African agriculture has become increasingly dualistic in character. By 1983, 30 per cent of farmers produced 75 per cent of all agricultural output. State policy has favoured this group, made up of rich individual farmers and of corporations, which now controls 80 per cent of the country's agricultural resources. But two out of every three farms are marginal and unproductive, and the rural economy as a whole has been sliding into a deepening recession.

Many of the poorer farms are in arid areas, where the ecology is very delicate, and are simply too small to be profitable. Serious over-grazing and over-cultivation have caused extensive erosion, making farms even less viable, and drought during the 1980s has made the problems even worse. Some small farmers no longer live on their farms, leaving black workers to manage them in their absence or to survive by subsistence farming. Poorer farmers have often opposed state action, such as the abolition of labour tenancy. On the other hand, some of the worst labour conditions are to be found on the poorest white farms.

Labour patterns among farm workers[6]

Despite the expulsion of some million farm workers and their families since the beginning of the 1960s, agriculture still provides employment for much of the economically active African population. Statistics are notoriously unreliable, however, and fail in particular to give any accurate indication of the numbers of casual, seasonal and part-time workers who are increasingly used on the farms.

Numbers employed and forms of work vary markedly with the type of agriculture undertaken in the various regions. While grain farming, which used to employ more labour than any other agricultural sector, has become highly mechanised and has laid off hundreds of thousands of employees, labour-intensive sectors such as fruit growing have expanded substantially, and have become the most important employers of labour.

Maize and wheat: Maize is grown in a number of areas, the most important being the 'maize triangle' of the Transvaal and Orange Free State highveld. Wheat is grown in a smaller area of the highveld and in Western Cape. A study of mechanisation on maize farms in Western Transvaal[7] showed a dramatic decline in employment. Between 1968 and 1981 the number of permanent workers fell by 20 per cent, and of seasonal workers by 50 per cent. Seasonal labour is increasingly drawn from the families of permanent workers rather than from the bantustans. As machines and chemicals make the work less physically arduous, so women and children on the farms – who can be paid less – are used.

In 1983, because of shrinking markets, the Maize Board launched a plan to reduce the area under maize cultivation and turn it over to cattle rearing, with an envisaged job loss of about 25,000.

Livestock: Livestock farms occupy 80 per cent of land in South Africa, but most are marginal. Mechanisation on larger cattle ranches has displaced large numbers of workers, but this has been counteracted by the growth of the feedlot industry and poultry farming, which are more labour-intensive than ranching. Corporate producers control a significant part of the poultry industry, and labour practices have begun to change as workers on factory-farms have become unionised.

Dairy cattle: Employment in the dairy industry has been reduced by mechanisation and has also become more like factory work, although unions have penetrated here less than in the poultry industry. Many small producers have been pushed out of the market, with more job losses.

Sheep: the Karoo – the arid interior of Cape province – is the centre of fine wool production. Farmers have not tended to invest in sophisticated shearing machines because of their reluctance to train workers, and because cheap hand shearers are easy to come by. However, as farms have become bigger and farmers' cooperatives have rationalised shearing, many jobs have been lost.

Plantation crops: Plantation agriculture, which is labour-intensive, has expanded rapidly in recent years. Crops include sub-tropical fruit (citrus, pineapple, bananas, tomatoes and others), cotton, sugar, tea, coffee and sisal. Much of the labour is seasonal and casual, recruited from the bantustans. The employment of women and children, for extremely low wages, is widespread.

Sugar: Grown in coastal Natal, parts of the Natal Midlands and Eastern Transvaal, sugar cane is essentially a labour-intensive crop, although mechanisation of post-harvesting processes has reduced labour requirements. Almost all cane-cutting is done by hand. Organisation among the workers in sugar mills has led employers to recognise unions on some plantations, but unions have been unable to organise on private farms.

A sugar plantation worker, Zululand. (Cedric Nunn/Afrapix)

Fruit and wine: Western Cape is the centre of the highly lucrative fruit and wine industry, which employs 10,000 whites and 140,000 blacks. Most farmers employ migrant workers for 11 months of the year, and wages are higher than in other sectors, owing to the Coloured Labour Preference Policy which applied until 1985. The industry's dependence on the export market makes it extremely vulnerable to international sanctions, which have helped unions to win some concessions for workers.

Forestry: This is another labour-intensive sector which is becoming increasingly important. Thirty per cent of plantations are state-owned, and 53 per cent of production is controlled by corporations. The Paper, Printing and Allied Workers' Union is making quite an impact on corporate-owned estates. Particular concerns are health and safety issues, such as controls on the use of chainsaws and herbicides.

A fruit farm labourer, Zululand. (Paul Weinberg/ Afrapix)

3 CONDITIONS ON THE FARMS

Working conditions on the farms vary considerably, but in general they are very poor. The semi-feudal relationship that prevails in some sectors of South African agriculture, particularly where forms of labour tenancy remain, means that whether or not workers are 'fairly' treated or mercilessly exploited depends almost exclusively on the good will or otherwise of the individual farmer. In some cases, and more in some provinces than others, farm workers have managed to win some improvements in wages and working conditions, with the help of unions or other support organisations. But such victories have been few and far between. Compared with industrial workers, who have made great advances in the last two decades, farm workers have hardly begun to win even the most basic rights and improvements. The main reason for this is their almost total lack of legal protection.

Outside the law[8]

In other sectors of the South African economy, workers are given some protection by the law. The Shops and Offices Act, the Factories Act, and the Mines Act all set down basic rules about working conditions for employees in these sectors. Four main employment laws protect industrial workers, but none of these applies to farm (or domestic) workers:

The Basic Conditions of Employment Act: This law lays down a maximum of nine and a quarter working hours per day for a five-day week, or eight hours for a six-day week, with statutory overtime payment for any extra hours worked. It fixes a holiday entitlement of at least two weeks a year on full pay, and establishes the right to certain public holidays and at least ten days' sick leave per year on full pay. It outlaws the employment of those under 15 years of age.

Farm workers usually work at least ten hours a day, six days a week, and at the busiest times of the agricultural year have to put in long extra hours without overtime pay. Many farm workers in fact work a seven-day week. Paid holiday leave and sick leave are both rare, and the employment of young children is common.

Labour Relations Act: This law lays down the ways in which workers and employers can negotiate wages and other working conditions. It gives workers the right to belong to a trade union and makes it an offence for the employer to dismiss workers for joining a union. It sets out procedures for reinstatement or compensation in the case of unfair dismissal.

Most farmers are strongly opposed to their workers joining a union, refuse union organisers the right to visit workers on the farms, and use the threat of dismissal to prevent workers becoming unionised.

Wage Act: Although this Act does not cover all industries, it establishes minimum basic wages for certain industries, according to area. In 1987, for example, the minimum for unskilled labourers in certain rural towns was R43 (£10.75) a week.

The majority of farm workers earn less than any basic wage set under the terms of the Wage Act. If a farmer provides food rations – and, even more, if he allows workers to grow their own food or keep animals on his farm – he will probably argue that wages do not need to be more than a small supplement. However, rations and workers' plots of land are frequently inadequate to provide for their needs.

Unemployment Insurance Act: This Act enshrines the right of most workers to unemployment payments for up to six months if they have worked in the same job for at least three years. It also establishes sickness and maternity benefits.

Farm workers (along with domestic workers, gardeners, seasonal workers and some government workers) have none of these entitlements. The only law which does cover farm workers is the Workmen's Compensation Act, which provides for financial compensation in the case of injury, sickness or death. Otherwise, farm workers can only fall back on the common law, under which any contracts they are able to enter into with their employer are legally binding.

Legal advice for farm workers

A group of lawyers and rural advice workers have compiled a two-volume handbook entitled *Your Guide to Farm Workers and the Law* for farm workers and those working on their behalf. Published by the Rural Legal Services Project and written in very simple English (Zulu, Xhosa, Sesotho and Afrikaans versions are forthcoming), it explains to farm workers how they can use the law, including common law, to maximum advantage. Among a wealth of other useful information, there are sections on courts and court procedures, pensions, compensation claims, how to delay evictions, and how to oppose unlawful impounding of animals.

346 pensions: payments

GETTING PAID

Where do you go to get the money?

Pensions are paid at different places in townships or in rural areas, for example a community hall, or magistrate's office or a shop or some other place.

Sometimes the place for payment is the same as the place for applications. But it is not always the same.

So you must ask the Pensions Officer where you must get your money.

When can you get your money?

From May 1989 all pensions are paid monthly.

But the dates and times of pension payments are different in different areas.

Your first payment can be bigger than the regular monthly payment, because the first payment is counted from the date of your application. So if you waited six months after applying, you should get six month's money in your first payment.

pensions: payments 347

How do you get your pension money?

- You must take your proof of identity with you on the day that you go for money. The pension number is stamped on this document.
- You must count the money to check that it is right and then sign or thumb print the card. This is proof that you have received the money.

If something is wrong, then you must not sign. You must complain immediately, to the person who makes the payouts.

Pensions advice from Your Guide to Farm Workers and the Law, *Book 2, pp. 346/7.*

Wages and payment in kind

Traditionally, farm workers have been paid in kind, rather than receiving a cash wage. The most common forms of non-cash payment have been food rations, fuel, and access to a plot of land on which to grow crops or keep livestock. Although payment in kind is still widespread, there has also been a marked shift towards cash wages, which means in practice that most workers are worse off than before, as the cash component rarely makes up for the accompanying loss of non-cash wages.

The forms of non-cash payment which have been most drastically reduced because of restructuring have been labourers' access to arable land, and permission to keep livestock. For the hundreds of thousands of workers evicted from the farms, particularly since the mid-1960s, the loss of such rights has of course been total. When they lived on the farms and had access to land, these workers were able to supplement the meagre and starchy rations provided by their employers with eggs, milk, meat and green vegetables. Eviction has often meant returning to the farms as migrant labour – but without any access to land.

Workers who remain on the farms have suffered too. Most have either lost all access to land or have been limited to small vegetable plots. If they do have access to farming land, it is much less than before. Workers commonly complain that the land to which they have access has been reduced from an area equivalent to three football fields to something roughly one-third that size. Only a small minority of workers remaining on the farms is allowed to keep livestock. A survey in 1976 revealed that even in the sheep and cattle rearing Karoo region only 40 per cent of workers had grazing rights, and a 1984 survey showed that only 19 per cent of workers on maize farms were allowed to keep livestock other than chickens or pigs. Again, those who are allowed to hold stock can keep only a restricted and decreasing number.

Of the traditional non-cash wages, only food rations remain widespread. These usually consist of a carbohydrate diet of maize flour, mealies (millet cobs), and beans. Among permanent workers resident on the farms, if a small meat ration is paid, it almost invariably goes to male workers. Most women workers and children receive no rations at all, or only a fraction of the amount given to men.

Cash wages vary widely between individual farms and between regions. Accurate figures are notoriously difficult to obtain, as official statistics are based on the farmers' own estimates of the combined value of cash and in-kind payments. This means that the value of in-kind payments is almost always over-estimated. Another problem is that farmers are asked to distinguish only between 'regular' and 'casual' workers, classifications which are too broad to show differentials between skilled and unskilled workers and between men, women, and children. The effect is that official statistics tend to inflate and obscure wage levels; and yet they still reveal appallingly low wages. In 1979, for example, the average monthly wage – in cash and kind – for African regular workers was R46

(£10.75) for the country as a whole. In regional terms this varied from R95 (£23.75) in the Cape Peninsula to R35 (£8.75) in northern Transvaal.

Wages can drop a long way below the average. A survey, conducted in 1981 by the Surplus People Project, of workers evicted from farms in the Orange Free State showed that 63 per cent earned less than R7 (£1.75) a month in cash. One of the most shocking cases was that of a worker who had worked twelve months a year for 24 years and was paid R3 (£0.75) a month in cash and nothing in kind, but had access to a small field. At the other end of the scale, wages of R500 (£125) per month are offered by some of the liberal farmers who are signatories of the Winelands Commitment (see chapter 5).

Farmers have at their disposal various other ways of 'reducing the cost of labour' – a euphemism for cheating the workers. Some pay part of the wage in the form of tokens, exchangeable for food and other provisions at the farm shop. Even when a token system is not used, workers often have little option about buying from farm shops, as they lack transport to reach other shops which may be many miles away.

The feudal master-servant relationship prevailing on most farms means farmers can withhold wages as a 'punishment' for alleged misdemeanours or poor work performance. Another way of effectively withholding wages is the practice of paying workers a so-called 'bonus' at the end of the season or at Christmas: in reality this means that workers are paid only a small part of their already tiny wage at regular intervals, and a lump sum after harvest or at the end of the calendar year. A 1984 study revealed that 58 per cent of permanent farm-workers received 80 per cent or more of their total cash wages as a deferred end-of-season payment.

The 'dop' system: alcoholic bondage

The winelands of the Western Cape are riddled with alcoholism. The infamous and archaic *dop* system, whereby employees are given a series of tots of rough wine throughout the working day, continues unchecked on about one in three farms, destroying the lives of workers and their families. Described as a 'perk' or a bonus 'ration' by some farmers, the *dop* is in reality a way of binding workers to the farms in a state of inebriated dependence.

Several commissions have recommended abolishing the *dop*, and the legal limit has been reduced over the years. A 1984 survey revealed that 13 out of 18 farms exceeded the current limit of 1.5 pints (just over a bottle) a day. Enforcement of the law is non-existent and tots amounting to a total of two or three bottles a day per worker are commonly doled out.

Although the *dop* is employed as a pernicious form of labour control, its natural consequence – drunkenness – is often severely punished. Nevertheless, many farmers expect a low-turn out for work on Mondays, as workers spend up to 20 per cent of their wages on weekend bingeing sessions in unlicensed shebeens.

The social consequences of the *dop* can be devastating. Alcohol-related diseases, domestic violence, child abuse and neglect are all commonplace. By their middle age many workers are worn out from the combined effects of alcohol and harsh labour conditions.

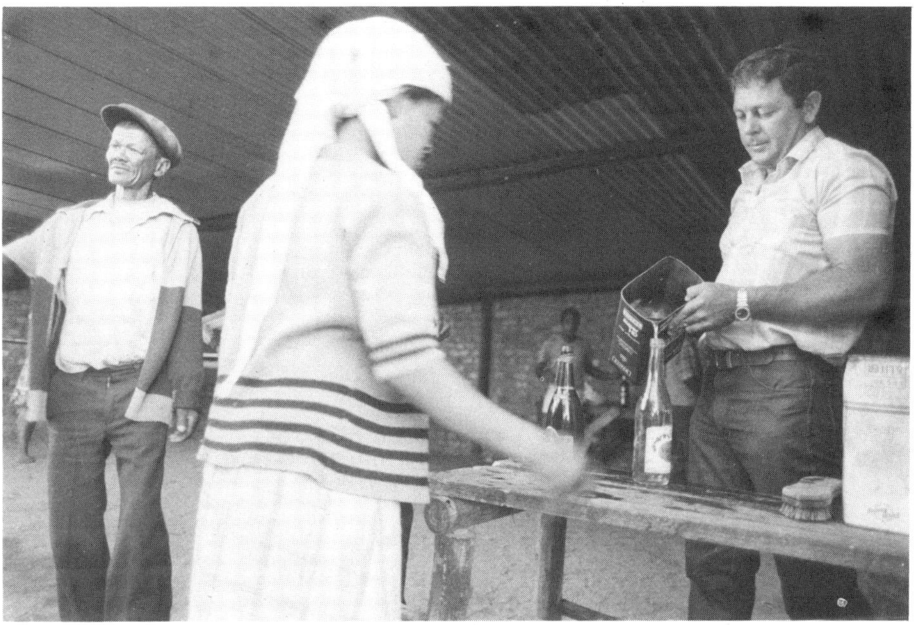

A farmer pours the weekend tot for his labourers on a farm at Vlottenburg, Easter 1986. (Orde Eliason/Link Picture Library)

Housing

Farmers often maintain that free housing is one of the perks enjoyed by farm workers. In reality, very few farmers provide housing, and when they do, it is usually grossly inadequate. Most workers build their own mud houses and are not provided with any services such as running water, sanitation or lighting.

Even the farm owners' South African Agricultural Union argues that there is an enormous need for better housing. A few farmers started providing rudimentary housing for workers in the 1960s and their numbers have slowly increased since then. With the restructuring of agriculture and the tendency to retain only a small number of 'useful' workers resident on the farms, the provision of 'decent housing' is seen as a way of stabilising the workforce round a reduced core of on-farm workers.

Although more farmers are building accommodation for some types of farm workers, changes in farming practice have on the whole enabled farmers to wash their hands of the responsibility to provide proper housing. Few make any provision for seasonal or casual workers, who are normally housed in sheds, outhouses or barns for weeks or months on end, or are obliged to construct their own shelters or sleep in the open. The practice of trucking workers to and from villages and towns in the bantustans is also increasing, and absolves farmers of the responsibility for providing any accommodation at all.

While some farmers acknowledge that some workers should be given adequate housing, they have generally been unwilling to pay for it. A state loan scheme exists to help farmers to provide workers' housing, but applications and grants are almost all for Coloured workers' housing in the Cape, very rarely for housing for Africans.

One type of accommodation fairly commonly provided is the compound, rather like a barracks, with shared toilets and washrooms – if these are provided at all. The Anti-Slavery Society gives the following example of the accommodation provided for ten young girls on a farm in the northern Transvaal:

> They lived in an L-shaped compound which had once been whitewashed and may have served as stables. Some of the windows were merely holes in the wall ... There were no toilet or water facilities apart from an outside cold water tap. There was no light, and fires had to be made on the floors in winter for both warmth and light. There was no furniture apart from cement beds built into the wall.

Health and safety

Not surprisingly, the general level of health among farm workers is very low. Overcrowding and poor hygiene and sanitation create ideal conditions for infectious diseases such as cholera, typhoid, polio and TB. Food rations are inadequate. Workers have little or no access to land on which to grow food, and wages are too low to buy nutritious food. Malnutrition, especially among women and children, is all too common. A report on farm schools[9] for the Diocese of Port Elizabeth in 1988 cites several cases:

> I was told by the teachers that the children have been starving since the 1983 drought. There is no adequate feeding scheme at the school, although food is sometimes received from the sisters at the King Convent. The clinic sisters have repeatedly noted severe malnutrition. The teachers say the children are lethargic and cannot stand up when they have reading lessons. In winter the lack of food is particularly pronounced, and the teachers say it leads to behavioural difficulties. Teachers also reported bad cases of gastro-enteritis, scabies and ringworm. (Smiling Valley, Cobongo)

> I was told that many of the children are losing their eyesight due to acute malnutrition. Stomach- and head-aches were also mentioned as a result of chronic underfeeding. An Operation Hunger school feeding scheme has been running at the school since 1987. (St Ignatius School, Addo)

> Malnutrition, hunger and related health problems were cited by all the teachers. (Sunday's River Valley)

Like so much else affecting farm workers, health care depends on the good will of their employers. Farmers often make scant – if any – provision for workers' needs. Sick leave is not guaranteed, so workers frequently continue working when they are ill, or else forfeit their wages. There is a dismal lack of rural clinics. Unless the employer is prepared to provide transport, there is often no means of obtaining medical help.

On farms where the *dop* system is practised, alcoholism brings grave health and social problems in its wake. It is not just adults who are affected. Children of women who drink heavily during pregnancy are often born with brain damage, or suffer from stunted growth. The rate of miscarriages and still births is high. Infant mortality in the Cape, at 64.4 per 1,000, is over three times the national average.

Occupational hazards on other types of farm are numerous. Farm workers can fall victim to animal-related diseases, known as zoonoses, which include anthrax glands and Malta fever. Exposure to the elements can cause heat stroke or Raymond's disease. Long hours, stress, monotony and insecurity give rise to nervous and mental disorders.

Mechanisation and the use of chemicals in agriculture expose farm workers to a variety of dangers, and at the same time they have less protection than any other group of workers against occupational diseases and accidents. Most accidents result from a lack of skill and knowledge, yet farmers are reluctant to provide any formal training – although there are substantial tax rebates on training expenses – for fear of employees leaving their service and seeking better paid jobs in the towns. In the 1980s about 2,000 cases of permanent disablement in accidents were reported every year.

366 common law health

If you are using any of these poisons on the farm, then you must stop immediately and first ask if the farmer has permission from the government. The farmer must have a paper to show that there is permission. The government does not give this permission easily. The poisons are always very dangerous, so the law usually says that they may not be used.

common law health 367

How to be careful

You have seen how important it is to wear protective clothes*.

Here are other ways to work with poison more safely:

Storage

- The place where the poisons are stored must be away from where people live or eat.
- Poison must be locked away.
- No children must be allowed in the place where the poison is kept.
- The containers must not leak, and all containers must have labels.
- Do not store poisons in coke bottles or other bottles that people may drink from.
- No smoking. Many poisons can catch fire or even explode like a bomb.

*PAGE 292-294

Pesticides advice from Your Guide to Farm Workers and the Law, *Book 1, pp. 366/7.*

Lethal substances used in agriculture have multiplied. Pesticides, fertilisers, fungicides, dipping chemicals, paint and fuel all pose health risks for workers, but few farmers supply protective clothing or training in the use of chemicals. The Machinery and Occupational Safety Act (MOSA) does include farm workers within its scope. This law requires farmers to provide protective clothing and instruction in how to work safely, to inform workers of potential dangers, and generally to ensure that working conditions are safe. But enforcement of the law is another matter. MOSA safety inspectors appointed to check working conditions often have no time to visit the farms, and when they do, they are usually more inclined to listen to the farmer than to undertake thorough inspections.

How many cases of poisoning from pesticides and other chemicals occur every year is far from clear. Department of Health statistics in 1981 noted 100 fatal and 92 non-fatal poisonings. The figures of the International Labour Organisation, on the other hand, are shockingly high: about 1,600 deaths every year from direct or indirect contact with pesticides and fungicides.

Annemarie Hendrikz, compiler of the handbook *Your Guide to Farm Workers and the Law*, explains why workers are particularly vulnerable:

> Many farm workers cannot read. They don't know what they are using. Pesticides banned elsewhere in the world have been used in South Africa, although there is very little information about this. There is also very little awareness amongst farmers about how dangerous these substances are. The whole environmental question has hardly entered people's consciousness in South Africa.

A crop sprayer on an avocado farm in northern Transvaal. (Steve Hilton-Barber/Afrapix)

> The dangers are simply everywhere, starting with the fact that workers don't know that they can become sick if they milk cows suffering from particular diseases or that if they don't handle a thrashing machine correctly it can cut off their feet. It's this whole system in which farm workers are treated like slaves. They just do what they're told to do, without any understanding of the work process, and this exposes them to untold dangers.[10]

In December 1987 the *Weekly Mail* reported on the use in Natal of one of the components of the infamous 'Agent Orange' (used in Vietnam with devastating consequences). A watchdog group, 'Chemwatch', based in Durban, has documented 300 cases of deformed children born to parents who were exposed to Agent Orange on the farms.

The Workmen's Compensation Act provides the means of claiming compensation in the case of work-related injury, sickness or death, but it has several loopholes and limitations. Casual workers (working fewer than three days a week) are excluded, as are prisoners (whose labour is often used on the farms). Often the farmer fails to fill in the claim form when a worker is injured. *Your Guide to Farm Workers and the Law* provides much useful information on how workers can set about claiming compensation.

Education

> There are few things more tragic than a child being robbed of its potential. Yet this is happening to hundreds of thousands of children. If there is no school on a farm or nearby, the children get no education. The State takes no responsibility for them.

The words of an educationalist at Witwatersrand University sum up the lack of schooling prospects for all too many farm children. Official figures state there were 5,627 farm schools for 481,325 children in 1988. Only one in three attends school at all, and the education available is usually of the most basic kind. Eight out of ten farm workers cannot read or write.

The provision of schools is entirely haphazard. None is constructed by the state. Instead, it is left to the farmer to choose whether or not to build a school on his land. If he agrees to build a school, he has to meet half the costs of construction himself. He then receives an annual maintenance grant, which falls far short of expenses. Some farmers dig into their own pockets to pay for the upkeep of schools, while many let buildings fall into disrepair.

Unless an outsider is appointed, the farmer is often the official school manager. He can close the school down at whim. He can – and frequently does – take children out of school to work on the farm during the busy times of year. Teachers are paid by the government but appointed and, if necessary, accommodated by the farmer, who therefore wields considerable influence over the content of the curriculum. There is a severe shortage of teachers willing to work in these beleaguered conditions, and those taking jobs on farm schools are usually poorly qualified.

School buildings are often cramped and uncomfortable, dark and badly ventilated. Hard concrete floors are freezing in winter and scorching in summer, and especially unpleasant for barefoot children. More often than not the bush serves as toilet. Furniture and equipment are insufficient. Children work two or three to a desk, sharing old text books or copying everything from the blackboard when there are no books. Most classrooms are notable for their unstimulating working environment – dull, dark, serviceable paint on the walls, a lack of visual materials and displays, no storybooks, atlases and reference books, no science equipment, no art and craft materials.

In many places children have to walk several kilometres to and from school. They are tired, and often hungry, before they even start the first lesson of the day. Absenteeism is common, caused by sickness or by the farmers' use of child labour. Most children never complete primary school, and for those who do, the chances of going on to secondary school are very slim. In 1987 there was only one farm secondary school in the entire country. To further their education, children have to board in the towns, an expense that ill-paid farm workers can rarely afford.

As a result, another generation emerges after a short school career, with minimal skills and doomed to a life of hard labour on the farms. This of course coincides precisely with the farmers' requirements. The whole system of education on the farms is designed not to open up opportunities, but to ensure a continuing supply of subservient labour.

However, workers are increasingly demanding better education for their children. The Orange Vaal General Workers Union found that workers put education as their highest priority for reform, even above wages and working conditions.

A primary school in Inanda township, Natal. (Omar Badsha/Afrapix)

School profile: Dwadwa

A researcher who visited farm schools for the Diocese of Port Elizabeth in 1988 reported[11] on a school where conditions are better than average.

Dwadwa farm school is in Lilyfontein, a fairly typical white-owned farming district. Farm labourers in this area were being paid an average R50 (£12.50) to R60 (£15) per month, with a 12.5 kg bag of mealie meal, and samp (maize porridge). They work from 6 a.m. to 6 p.m. mostly, five days a week, and from 6 a.m. to 1 p.m. on Saturdays. Most of the parents have had very little formal education, as from the age of about twelve the children are expected to work on the farms. The farmers, mostly of German origin, are very conservative. They not only offer very little support to the schools (with one or two exceptions) but are often overtly hostile. Many of them resist the idea of children going beyond Standard 4 (about six years of schooling). During the busy times on the farms children are expected to work, thus interrupting their schooling. The long hours worked by the parents and their low wages mean they can give little effective support to the school, despite their willingness to do so. Workers often have to move from job to job in the same area, and this migratory pattern is highly disruptive of the children's schooling.

Dwadwa school in 1987 had 115 pupils. The school is housed in two wattle and daub structures. Pupils spend much of their time maintaining the buildings, particularly after the rains. The rooms are very dark and inadequately ventilated. The roof leaks constantly when it rains. However, the buildings are little short of a miracle, considering they were erected by two not young women (the teachers) and kindergarten and junior primary pupils.

Our impression of the school was that, although extremely poor and neglected, it was a happy place filled with a lot of love and spontaneous activity.

Some of the most pressing problems raised were:

Premises and accommodation: The school is in most urgent need of four classrooms, toilets, reliable water supplies, a storeroom and a staffroom.

Young pre-school children: A group of pre-school children come to school with the older children as they are unable to stay at home alone all day. Their presence makes the teaching/learning process even more difficult. It is also detrimental to the little ones' development to demand the standard of behaviour necessary for formal classroom-based learning.

Parental involvement: The pressure of work on the farms limits parents' and guardians' involvement in the school. However, they are keen, and were in the process of forming a parents' committee. Parents often ask the teachers to write letters for them, and for advice regarding a variety of problems.

Educational and support needs: The two teachers are comparatively well qualified and are extremely dedicated, but they lack professional support. They have been teaching in almost total isolation for 15 years. They lack basic teaching aids such as charts, maps, reference books, picture books and crayons. The Department of Education provides only the most basic of items.

Church literacy scheme for adults

Southwell Anglican Mission in the Albany District of the Eastern Cape (funded by Oxfam) has become deeply involved in outreach to farm workers and, in particular, in running adult literacy and English classes.

At the night school, farm workers made their own statements about their lives and education:

> The pain in my life is this: my parents didn't send me to school. Now, if I can go to school, I will never give up. When I asked my father about school, he said to me, 'Who is going to look after the cows?' If I can have a little education, I will be very glad. With regard to education, I am in a sad plight, because now my eyes do not see very well. ...

> The reason that I never studied is because I had to start work when I was little. My mother was not healthy, and my father died. That's why I didn't learn at school. Even these few words I've had to request someone to write for me. All my life I have wanted to learn. ...

> The problem is this: when I was a schoolchild my father got sick and he went to hospital. He died when I was still at school. I was left with my mother. My mother got TB, and then she went to a TB hospital. She didn't have money, so I had to leave school. Although there was no money to go to school, the Lord was at my side. I was able to come to night-school. ...

> My story is I leave school in 1964 and I get a job at the fields. I was getting 75c a week. I was seeing my home getting more and more difficulties. If I was educated I could be the good helper of my home. Now I was married and my husband never went to school. That is why I want even night-school.

Seasonal and casual labour

The most drastic consequence of the mechanisation and restructuring of South African agriculture has been the destabilisation of the work force. A small core group of workers has acquired new skills as machine operators and supervisors, and since farmers have to pay higher wages for skilled work, they have sought to reduce the cost of unskilled labour. The answer has been to rely more and more on seasonal or 'part-time' labour.

Here we enter a word game designed to obscure the true nature of workers' employment. Full-time agricultural workers are described by the commercial sector as 'permanent' or 'regular', and part-time workers as 'seasonal' or 'casual'. All the terms are misleading, used subjectively by the employers. Migrant workers employed on contract are designated 'part-time' although they may 'commute' to the farm all year round. Dairy parlour workers doing 'short' shifts are also classed as 'part-time', although they may work two shifts a day, each of several hours' duration, seven days a week. Workers who are full-time but paid for piece-work are also deemed to be 'part-time'.

In general, employment has become casualised and the status of workers has changed from 'servant' to 'migrant'. In addition, the work process itself has been reorganised, broken down into tasks which can be allocated on a piecemeal basis. For example, maize harvesting is divided into various components, including reaping, threshing, transporting, and storing. This breaking down of tasks is seen as an essential part of 'modern farm management'. It enables farmers to control the labour of 'full-time' workers more strictly and use their labour time more intensively, as well as allocating defined tasks to 'part-time' workers.

Much of the migrant labour employed in labour-intensive sectors of agriculture is drawn from the bantustans, where former on-farm workers, evicted by white farmers, live in desperately overcrowded villages with no prospects of local employment. Thus 'full-time' workers return to the farms as 'part-time' workers, earning lower wages for a longer working day (when the time spent travelling is taken into account). Often it is only women and children who are desperate enough to accept such exploitative conditions.

Many farmers send open trucks into the bantustans to pick up workers on a daily basis. Such 'commuting' adds hours to an already over-long day. Travelling in open trucks is uncomfortable, cramped and dangerous – accidents are frequent – and workers are either made to pay for each trip, or they pay indirectly if the provision of transport is treated as payment in kind. If travel costs are too high, farmers may opt to accommodate workers in barns or sheds during the week and

Eastern Transvaal: Mr Timothy Hlatswayo, a farm labourer who was injured while being transported to work. (Gill de Vlieg/Afrapix)

drive them back to the reserves at the weekend. In some cases farmers provide transport to work in the morning, but not back home in the evening. In February 1990 an Oxfam field officer reported seeing workers tramping ten miles back to villages in Venda at the end of a day's work on plantations in the Transvaal.

For others the journey is much longer. Migrant workers from Taung, part of Bophuthatswana, face journeys of over 500 km to farms in the Transvaal, where they work for periods of 3-6 months. In 1979 two teenagers hired as harvesters froze to death on the back of a lorry which was driving through a blizzard in the Orange Free State. For such migrant workers the average cash wage for 1988/89 was R2.90 (£0.72) for a 12-14 hour day. Payment in kind varied between 4.5 and 11 bags of mealies per 100 harvested.[12]

Perhaps most tragic of all is the case of refugees who have fled from the war in Mozambique. They are permitted to stay in the homelands, but they are not allowed into 'white' South Africa. Poverty and unemployment are so acute in the homelands, however, that many refugees have run the risk of seeking work on white-owned farms in north-east Transvaal. In July 1989 local relief workers reported that a white farmer had recently hired 30 Mozambican refugees staying in Gazankulu to work for him for three weeks. The refugees agreed to work for R3 (£0.75) per twelve-hour day. They received their wages after the first week, but nothing at the end of the second week. At the end of the third week, instead of paying them, the farmer reported them to the South African Defence Force as illegal aliens, whereupon they were immediately deported back to Mozambique.

Another source of seasonal labour is, increasingly, families and dependants of male workers resident on the farms. This is particularly the case on semi-mechanised farms where the number of full-time male employees has been reduced.

Women's work

Women on the farms are even more vulnerable than men. Most permanent farm workers are men, and women usually have the right to live and work on the farms only through their attachment to male labourers. A pre-condition of the male worker's employment is often that his wife will work as a domestic servant for the farmer, or be available for seasonal work whenever required. In some cases this work is unpaid. If they are not employed on the farm, women may succeed in finding low-paid jobs – usually as domestics – in nearby towns, but the travelling involved makes extra demands on time and energy.

In addition to their work on or off the farm, women bear the responsibility for cooking and cleaning, and for looking after children, the elderly and the sick. Strenuous tasks like fetching firewood, or carrying water from the stream or water pump, all fall to women. If access to a plot of land forms part of the male worker's terms of employment, the woman's work burden is further increased by having to tend animals or help cultivate food crops on her husband's fields. Women's position on the farms is even more insecure than that of men. If her husband dies, a woman almost invariably loses her home.

The shift away from permanent and towards casual labour on the farms has also meant increased female employment. Full-time jobs have been broken down into 'part-time' components, and dirt-cheap female labour recruited to carry out the various tasks as and when necessary. One farmer's satisfaction speaks for many others: 'We are making more and more use, with very pleasant results, of the women doing the work which was formerly the accepted work of the men.'

Farmers often attribute these 'very pleasant results' to women being more 'meticulous' and 'efficient' than men. This is a cover-up for the real reason for their employment: the fact that women are invariably paid even less than men. Piece-work is common, and bad weather means no wages. Casual labourers are rarely given rations or protective clothing either, so they are more prone to ill-health and injury.

A myth has grown up that 'part-time', casual work, especially where mechanisation has taken place, is by definition 'light work'. Sociologist Tessa Marcus points out that most women are in fact engaged in heavy, physical labour:

> They spend from sun-up to sun-down in the fields on back-breaking tasks, plucking fruit, tea, tobacco and cotton; lifting chicory, groundnuts, potatoes and cabbage; picking tomatoes, pineapples and flowers, etc. They glean behind the combine harvesters which, fitted with overhead lights, move through thousands of hectares of maize and wheat day and night. They wield hoes the year round, weeding the fields. They thin the grapes, lay and relay the irrigation pipes, build and break down the banks of flood-irrigation canals, milk the cows, churn the cream, wash down the milking equipment and the sheds, herd the livestock.

Sexual exploitation is widespread. With rising unemployment, the only way women can get or keep jobs may be to grant sexual favours, or to file no complaints in the case of favours taken, or of actual rape. At home, too, women may suffer sexual abuse or physical violence from their own men who, ill-treated themselves by oppressive employers, take out their frustration on the women.

Anna's story

Anna Mazibuko told her story in 1984 to a worker from Black Sash.[13] (Black Sash is a South African women's organisation which works to promote justice and human rights. Oxfam has funded its advice office programme, and has helped to pay the salaries of rural fieldworkers.) Anna's story typifies the rootless existence of many black women in white farming areas, alternating between farm work and domestic service.

> The first place I worked was on the farms. I worked in Amesfoort (South Eastern Transvaal) in the yard of Jan Niekerk. I was 12 years old and my job was to look after the children of the *nonantjie* – his wife. But at that time I didn't earn any money. I just worked for a plate of food.

Then my family moved to Morgenson in the Eastern Transvaal. It is near Standerton. I was 14 then. I worked on a farm and we used to plough there. We didn't earn wages then. That was the time when you got some of the crop, but there was no crop that year, so we moved after six months. There was no crop because there was no rain.

Anna ran away from the farms and took a succession of domestic jobs in towns, before joining her sister on a farm in Holmdene (Eastern Transvaal). She was paid R4 (£1) a month for weeding, threshing and binding mealies. After a spell back in the towns as a domestic worker, running away from several jobs because of harsh employers, Anna returned home.

When I went home I had a baby. She only lived for one year and a month. She died. So I left. I decided it was the time to get married, so I set off to find a man. The one I found was no good, but I had many of his children. I had nine, but five died.

That man was from Standerton. He's from the farms, and he's there until today. I couldn't agree with him so I left. Last week I went to look for my children. I found the youngest who is ten. The only one I have with me is my daughter. The three boys are with him. I am separated now from that man. But I said to him he should keep the children, because I knew when I left I'd be very poor.

Life on the farms is very hard. It's very heavy. In Standerton where I lived with the father of my children we would get up at six in the morning. The woman of the family (that is I) would have to start cooking and the man would have to start milking. After cooking for the children and my husband, I would go to the farmhouse and do the cooking and cleaning. I earned R12 (£3) a month.

At the farms there's a lot of ploughing. If it does not rain you're in a lot of trouble. My husband has always earned R30 (£7.50) a month and a bag of meal. And he would get the corners of the fields (the *agterskot*), after the reaping. So if the crop died you died.

The farmer never gave us extra. And yet he knew my husband was a man with a wife and children. In the years when there was no rain, those were the years when our children died.

Even now that man is still milking at six o'clock. Every afternoon at 4 p.m. he calls them in again to the shed and sits down beneath them and starts again to milk. That's the law of the whites. Whatever the weather, wherever you are. There you are under the cows.

In the morning after the milking he must go out and plough the fields on the tractor, skoffel (weed and hoe the fields), start again to plough the fields ... and at four o'clock he must get off the tractor and get under the cows.

Next morning same thing. Then when you've finished ploughing and skoffeling, back to the cows. And after you've milked the cows you must go to make cream on the machine. And after the cream he can go home. That was his life. Day in and day out.

As for me, my life was a white lady's kitchen – day in and day out. All for that R12 (£3). My husband gave me R10 (£2.50) a month. I was supposed to buy everything for my children and myself. I couldn't manage. So I said good-bye.

Working in the white lady's kitchen. (Afrapix)

No child's play

According to Anna's story, she began working on the farms at the age of twelve. This is by no means unusual. Child labour has been used in South African agriculture since the early days of European settlement, and continues to be widespread in 'modern' South Africa. In 1978 the International Labour Organisation estimated that 65,000 children between the ages of eight and fourteen were used as farm labourers.

Restructuring and the increased use of casual labour have made the employment of under-age children particularly popular with farmers. Not only is their labour even cheaper than that of adult males, but children are more malleable and can be more easily exploited. Children are recruited from the bantustans, and to a lesser extent from townships in the Republic, where poverty and hunger make them easy targets. Along with women, truckloads of children are transported to and from the farms, either travelling each day in addition to the long hours spent working, or being housed in rough accommodation during the week and returning home for all or part of the weekend.

In the early 1980s the Anti-Slavery Society conducted research into the extent of child labour in South Africa. As most working children in the Republic are found in agriculture, the Society carried out numerous interviews in the white farming areas, talking to children whenever they were able to gain access to them, and listening to the testimonies of former child labourers.

In the Eastern Transvaal the Society's researchers encountered children between the ages of seven and 15, working in maize fields and herding cattle and goats. Some lived locally and had left school to work on the farms, while others came from the bantustans. Many of the children claimed to have been misled about the nature of the work expected of them. As one girl working on a poultry farm said, 'We were tricked. We were told we would look after chickens, but when we got here, we were given work in the fields.'[14]

The Society found that children were often recruited without the knowledge or consent of their parents and were rarely registered by the farmer. They received below-subsistence wages and lived in poor unhealthy conditions. If they fell ill or had an accident, no one could be held responsible.

In Natal, remaining labour tenancy arrangements often depend on providing child labour. The Society found many cases of children working on behalf of their families. Sometimes the whole family was required to work for the farmer. Sometimes it was only the children who were taken on as labourers. Some tenants employed children from other families to work on the farms, while they were employed elsewhere.

Child farm labourer near Magaliesburg, Transvaal. (Paul Weinberg/Afrapix)

Numerous other cases of child labour were found – in the fruit farms of the Cape, on sugar plantations in coastal Natal, in vegetable and cereal production, and in animal husbandry.

A particularly distressing aspect of the employment of children in agriculture is the widespread sexual exploitation of young girls. One researcher points out that girls are often specifically recruited because they are female, and therefore they not only suffer as child labourers but 'bear the full burden of womanhood too'.[15] Often girls are expected to provide sexual services to the male workers, as well as to the farmers.

Ten hours a day for 33 cents

Seven-year-old Lucas Makgokolo is being exploited by his 'boss' as a casual farm labourer on a farm in the Vaal Triangle, but he is not old enough to join a union or claim compensation for injuries he recently received while at work. As sheep herders, Lucas and his brother Ismael (12) each bring home a wage of R2 (£0.50) a week.

When Lucas turned six, the farmer asked his parents to allow him to work for him full-time. They refused and said he had to go to school, but since then Lucas and his brother have had to work from 7 a.m. to 5 p.m. during school holidays. The wage works out at fractionally more than 33 cents a day.

Last year Lucas was bitten by his employer's dog and as a result is believed to be suffering from brain damage. His parents approached the Orange Vaal and General Workers' Union (OVGWU), who are now involved in a legal battle to claim compensation for Lucas's injury.

Although the Employment Act basically prohibits the employment of children under 15, Lucas is just one of thousands of farm children in rural and semi-rural areas who are in the front line of the workforce of farm labourers.

A researcher for the International Labour Organisation said she found some of the worst conditions in the Vaal triangle. 'One particular farmer who had a farm school on his property forced the teacher to act as a foreman and threatened to close the school when the teacher resisted.

'Many parents who are farm labourers are forced to consent to their children's employment because they fear being fired,' she said. Others, like Mr Abraham Moekoena, who works on a farm in the Vischgat, Vereeniging area, consent to their children working because the family could not survive on his meagre wage of R40 (£10) a month and a bag of mealie meal:

> The farmer employs 36 adults. The 30 children on his farm aged between eight and 18 are all expected to rise at sunrise and work till sunset during the school holidays and on Saturdays. The farmer rounds up all the children and they work under an adult foreman reaping mealies. I myself began working on a farm at the age of eight and I am now 42. I don't believe children should work – they should have time and food in their stomachs for their school work. But without the wages they bring in, we would all starve.

In 1986 the Department of Education and Training claimed it was unaware that some white farmers were closing schools during harvest time and forcing pupils to work without pay. In January 1988 the Government announced its intention to ban the use of black children at farm schools from any form of labour during school hours.

In terms of the education laws, anyone who removes a child from a farm school during school hours can be fined up to R150 (£37.50) or jailed for a month. But so far there have been no convictions. As the ILO researcher pointed out, there is at present no monitoring of the child labour situation in South Africa.[16]

Prison labour

In 1989 a journalist from the *Weekly Mail* went to Leeuwkop, a prison on the outskirts of Johannesburg, and 'bought' two convicts for the day for the sum of R2.80 (£0.70). The visit confirmed reports that the system of selling convict labour for use on the farms and elsewhere was still in operation, despite assurances by the government three years earlier that it was to be phased out.

The use of prison labour violates international agreements to which South Africa is a signatory, including the General Agreement on Trade and Tariffs (GATT) and the codes drawn up by the International Labour Organisation. Yet the prison labour system remains widely practised in agriculture, as well as in certain other sectors of the economy.

Not all farmers use prison labour, and some are strongly opposed to it. For those who do use it, there are two main sources: the parole system, and a lease scheme, commonly known as the 'buying of bandits'. In both cases prisoners' labour is almost invariably forced. Theoretically, parole prisoners are given a choice between remaining in prison or working on the farms for 'market wages'; in practice they are often given no choice at all about going on parole.

Under the lease scheme farmers can pick up exceedingly cheap unskilled labour on a daily basis, the only requirement being that they swear to act as temporary warders while convicts are working on their land. Although farmers have to agree to prevent convicts from being assaulted, there is no check on the farmers' conduct, and complaints about ill-treatment to the prison authorities by the prisoners themselves normally fall on deaf ears. On the other hand, if farmers are dissatisfied with a prisoner's work, a complaint to the warders is likely to result in a spell in solitary confinement – a further element of compulsion in the prison labour system.

Prisoners working on the farms are frequent victims of violence and abuse. Yet their ill-treatment cannot really be separated from that suffered by other farm workers. The control exercised by farmers over their employees is so great that the distinction between voluntary and forced labour is often very blurred.

Violence against farm workers

On 23 February 1990, the *Sowetan* newspaper ran two front-page stories. One, referring to the President, F.W. de Klerk, was headed 'FW agrees to meet the ANC'; the other, under the headline 'Farm death row', concerned the case of a Piet Retief farmer, fined R100 (£25) for assaulting his employee, who later died. The juxtaposition of these stories – progress towards justice and an example of flagrant injustice – reflects the contradictions of South Africa in the 1990s.

The assault by the Piet Retief farmer took place several months earlier, in August 1989. Ekathi Xaba, a partly paralysed farm worker of 37, was attacked by his employer, Wilhelm Herman Rabe, for staying away from work, and died the next day. It was not until November that Rabe was arrested. Two key witnesses, a fellow farm worker and Xaba's 16-year-old daughter, who had seen the victim kicked, beaten and trampled on, were not called to give evidence. The magistrate rejected any link between the cause of death – a subdural haemorrhage of the brain and a ruptured spleen – and the hour-long assault. He acquitted the farmer of culpable homicide and found him guilty of common assault, fining him R100 (£25).

Xaba had been employed as a wood sorter on Rabe's farm. A few years earlier he had been partly paralysed after a tractor had run him over on his previous employer's farm, but he received no compensation for the injury. After his death, his widow was driven off the farm with her children.

Physical assault and abuse are an inseparable part of the paternalistic system that binds workers to the farms and allows farmers excessive control over them. Beyond the 'routine' violence prevalent on so many farms, there are also chilling cases of serious assault, torture and murder.

Simeon Mambakazi worked on a farm near Piet Retief for 25 years until one day his 'master' chained him by the neck to a tree in the forest, leaving him there for eight hours in an attempt to make him confess to stealing zinc sheeting, which he denied. Eventually the farmer freed him, but Mambakazi's neck was so swollen that for a week he could eat virtually nothing.

Stephen Monyone, a 28-year-old farm worker, was accused by two Western Transvaal farmers of stealing two cows. They tied his hands behind his back, dumped him on the floor and mercilessly kicked him in the stomach and beat him with sjamboks (clubs). He died the next morning.

Mfanufikile Henry Zondi, a pupil at a secondary school in Wartburg, Natal, was walking home from school along the tar road leading to the Windy Hill Wattle Company farm when he was accosted by two black security guards from the farm. Apparently, farm labourers had been told not to use this road, although there was no other road in the area. One of the guards blew his whistle and the farm supervisor allegedly struck Zondi on the head with a brick. He was knocked unconscious. When he revived, he was taken and handcuffed to an outside post and not released until the following day. When he returned home gravely ill, his mother called an ambulance, but on arrival at the hospital Zondi was declared dead.

With the help of the Association for Rural Advancement (AFRA) (funded by Oxfam), the family succeeded in taking the farm supervisor to court. Two women who had earlier made statements to the police about the assault failed to appear as witnesses. The defendant denied hitting Zondi with a brick. He said he punched him in the face, knocking him unconscious for a brief period. He was sentenced to a fine of R50 (£12.50) or 20 days' imprisonment for assault, and to R600 (£150) or nine months' imprisonment for kidnapping.

AFRA suggests a number of reasons why assaults continue unabated in the farms. For one thing, farm workers are isolated and unaware of their rights. For another, their position in general is extremely insecure, and they are afraid of the consequences of reporting such cases. Even when cases are reported, they are often not followed through systematically, and few prosecutions take place. And even where people are prosecuted, the delay in investigating often means that witnesses fail to appear before the court, and thus the accused are often acquitted for lack of evidence.

The incidence of farm brutality appears to have been rising in recent years. Researchers have suggested various reasons, including increasing insecurity among farmers whose state subsidies have been reduced or withdrawn because of economic recession, a resurgence of right-wing extremism among farmers, and prejudice and corruption on the part of the police and judiciary.

4 EVICTIONS

Farm workers living on the farms are, as we have seen, locked into a tightly controlled, highly exploitative system. Yet life on the farm, however harsh, is seen as preferable to the harrowing alternative of losing home and livelihood. The threat of eviction, whether actual or potential, is for many farm workers the greatest fear of all.

Although the massive, state-sponsored evictions of the 1960s and 1970s are a thing of the past, evictions continue today almost unremittingly, even if in a more piecemeal fashion and on a smaller scale than before. Increasingly, they have been carried out as the result of private action – with state support – by farmers and agricultural companies against workers who are no longer wanted on the farms.

Workers are in danger of dismissal and eviction for all sorts of reasons. With increased mechanisation and the concentration of larger tracts of farming land in fewer hands, farmers have been streamlining and reducing their permanent workforce, often replacing permanent workers with seasonal and casual workers. When farms change hands, new owners often feel no obligation towards farm workers whom they 'inherit' from the previous owners. On farms where some form of tenancy arrangement still exists, the failure by tenant families to replace ageing workers with younger, fitter members of the family can often result in notice to quit. Workers who complain about poor conditions or attempt to join a union are frequently dismissed for being 'trouble makers'. In general, the all-powerful position of the farmer as employer and landowner enables him to take callous or punitive action against workers if he so desires – expulsion being the ultimate sanction at his disposal.

Prevention of Illegal Squatting Amendment Act

Only two years after the Abolition of Influx Control Act of 1986, the government introduced a bill which effectively brought back influx controls, but in a very much more severe form. Despite a sustained campaign against it by democratic organisations, big business and parliamentary opposition, the notorious Prevention of Illegal Squatting Amendment Bill (amending the Act of 1951) became law on 9 February 1989. As yet there have been no prosecutions under the Act, but it is often used by farmers as an excuse to escalate evictions.

The Act has had very serious implications for farm workers. Under the 1951 Act it was an offence to enter or remain on any land without the owner's permission. If a person was charged with such an offence, the State had to prove his or her guilt. With the new Act, the burden of proof has passed to the accused. The penalty for disobeying the legislation has been increased to R10,000 (£2,500). A magistrate has no discretion about judgement, but has to order the 'squatter' to leave his or her house. Shacks may be demolished without a court order, and the courts cannot stop a demolition once it has begun, even if it is unlawful. Only those people who work on farms are allowed to live on the land, which means that the family and dependants of workers are liable to eviction. A landowner who allows anyone other than workers to live on his farm is liable to a R10,000 fine.

Oxfam partners such as AFRA (the Association for Rural Advancement), TRAC (Transvaal Rural Action Committee), and the Black Sash all testify to the punitive effect of the new Act. Farmers wishing to eject unwanted workers and their families now find it even easier to do so, while those who allow 'non-workers' to remain on the farms are breaking the law and risking a heavy penalty.

The case of the Kei Road Community

A struggle on behalf of 36 people evicted from a farm in the Kei Road area of Eastern Cape in February 1990 illustrates the vulnerability of farm workers' communities, and the constraints imposed on groups trying to help them. Unusually, it was neighbouring white farmers, Laura and Tony Goodyer, members of the Anglican church, who rose to the evicted workers' defence and enlisted the support of a number of other church people including the local priest, the Reverend Martin Glover, the Bishop of Grahamstown, and Ntombazana Botha, a social worker who had been working with the Border Council of Churches.

The eviction arose out of typical circumstances. A farm which had been home to four families for more than 20 years changed hands. The previous owner, Milton Adler, was considered to have cared for the farm workers. Wages were

Anxious times for members of the Kei Road farm workers' community, March 1990. (Gill de Vlieg/Afrapix)

poor (R44 a month – £11), but they had plenty of food and were allowed to keep cattle, goats and poultry and use the tractor for ploughing. Adler regularly provided extra milk and dispensed drugs when people were sick. When a baby was born, he supplied baby food and extra provisions, and when someone died, he would provide a sheep and groceries for the funeral wake. Rather than make people carry firewood on their heads, he would allow them to use the tractor to collect it.

After Milton Adler's death in March 1989 a neighbouring farmer, Sid Knott, was appointed manager of the farm and later bought it. Wages were paid erratically and ceased altogether when the workers and their families were told in September to leave the farm by 1 February 1990.

A self-defence action group was formed, consisting of the Reverend Martin Glover, Tony Goodyer, and two representatives of the farm workers, Alson Tshangela and Fashion Somtsora. Bishop David Russell of the Diocese of Grahamstown visited Sid Knott 'to listen and plead as a brother Christian', since Knott was also a member of the Anglican church. The local press was contacted and gave extensive coverage to the affair. But, in the end, all the forces mobilised in support of the farm workers proved powerless to stop their eviction. The Goodyers' action was bitterly resented by local farmers, who severely harassed them.

After the eviction the Kei Road Farmers' Association held a meeting at which they condemned the church's 'meddling' in what they saw as a routine farming practice. They described as 'ridiculous' well-publicised comments by the Bishop about farmers' exploitation of their workforce. A claim by the farmers that Bishop Russell, the Reverend Glover, and Tony Goodyer had not had the courage to attend the meeting turned out to be a distortion of the truth. They had in fact welcomed the opportunity to meet with the farmers, but on condition that farm workers' representatives were also present. They argued that it was no longer the practice in South Africa for white groups to talk about what should happen to blacks without their being present.

Despite the failure to stay Sid Knott's hand, there were many achievements both before and since the eviction, which should not be underrated. A significant number of people provided practical help: a local doctor undertook a complete medical examination of the community at their request; financial support was given by the South African Council of Churches and Bishop David Russell; the Legal Resources Centre in Grahamstown provided legal advice; Ntombazana Botha recorded the detailed history of the evicted farm workers on tape and transcribed and translated the interviews. Several Anglican clergy visited the community to offer moral and spiritual support.

At first the evicted families were accommodated on the Goodyers' farm in tents provided by the SACC. Meetings between the farm workers and a representative of the Cape Provincial Administration (CPA) led to an agreement that the whole community would move to the nearby township, where tents would be provided

Kei Road children settling into their new home, March 1990. (Gill de Vlieg/Afrapix)

by the CPA, until a more permanent alternative could be found. It was also agreed that any other farm dwellers similarly evicted from Kei Road farms would be free to join this community, and that tents would be made available to them too.

At the time of the community's move to the township, Tony Goodyer pointed out the significance of these achievements. First, the Kei Road workers had been brought together as a united and organised group. Second, they had been recognised as *bona fide* farm workers and not squatters, as the farmers claimed. Third, the allocation of land and tenting, albeit temporary, provided a rallying point for other people evicted from local farms and struggling to find shelter. The example of the Kei Road community may well inspire others elsewhere to realise that they too could make a difference to their lives by taking similar action, so promoting their fight for legal rights.

The story of the Kei Road community is not finished yet. It had been hoped that the farm workers and their families could move into St John's, a nearby farm owned by the church, but this provoked threats of retaliation by surrounding farmers. So in July 1990 the church land trustees sold St John's to buy a larger farm which, because it borders on the Ciskei, was available at a good price. The land there is fertile, with good irrigation and storage facilities. On the new farm (to be called St John's Hillside), the farm workers plan to form themselves into a cooperative. They have been offered agricultural training by a local organisation, and Oxfam is likely to fund the purchase of 150 Duni merino sheep, a highly valued breed found only in that area, and jealously guarded by the white farmers.

Still together: the Kei Road community encamped near Emthonjeni township, Eastern Cape, March 1990. (Gill de Vlieg/Afrapix)

Labour tenants: struggles and resistance[17]

In areas such as South Eastern Transvaal, Northern Natal, the Natal Midlands and parts of the Orange Free State, where variations on the labour tenancy system remain, people are struggling to hold on to the land where their families have lived for generations. In the face of eviction orders, trespass notices, court cases, fines, demolition of their homes, impounding of livestock, and even prison sentences, most tenant families show dogged determination and resistance in their attempts to hang on to their homes and way of life. There are families who have been evicted many times and have returned each time to the same farms. One old man, Sergeant Yende, from a farm near Driefontein, was evicted eight times, and returned eight times to his homestead.

What accounts for such tenacity in clinging to the land? Why are evictions so fiercely resisted? The explanation, according to former TRAC fieldworker, Aninka Claassens, is that although in terms of cash income these people are extremely poor, they often possess a kind of wealth denied to most black people in South Africa. Their agricultural capital can be quite substantial: some families have large numbers of cattle, sheep, goats or chickens, often of excellent stock, and have acquired farm machinery. Their homes, which they may have improved and enlarged over the years, can accommodate the extended family, and from this family structure tenants are able to draw support and a sense of self-worth which enriches their lives.

Eviction puts all these 'assets' at risk. Animals and machinery cannot be taken to the overcrowded townships or relocation areas, and local white farmers often collaborate to prevent them selling for their true value; space and a valuable measure of independence is lost; the extended family is broken up.

Fieldworkers also point to the fierce pride of these people in asserting their inalienable right to their homes. Questions about who owns the farms often meet with disdainful replies:

> The whites have not been here for long. This one has been here for six months. The one before was just the same ... We are the people who live on the farms. We are not people who are visiting or people who are passing through. We were *born* here. Another place we have not seen.[18]

This conviction of their right to be there helps sustain tenants in their struggle against eviction. Yet the isolation of these families, the usually racist nature of the police, judiciary and other institutions, and the legislation relating to 'squatting' and 'trespassing' all stack the odds against them. In tightly-knit rural communities, the police and magistrate may often be related to the farmer responsible for evictions. (Historically, the local 'veldkornet' – Justice of the Peace – was elected by the farmers to be responsible for justice and administration of the law.)

Nevertheless, with the help of service organisations such as TRAC and AFRA, and advice offices and para-legal clinics run by organisations like the Legal Resources Centres, some legal cases have been won. In most instances, the only basis for challenging eviction is the shortness of the notice period or the unreasonable way in which a contract is terminated. The recent victory of Johannes Mtjali and Kleinbooi Nene, widely reported in the South African press, is an example of a successful challenge to eviction. The magistrate ruled that one month's notice was totally inadequate for the two defendants, both aged 74, who had laboured for decades on the same Transvaal farm without receiving any wages. Yet, as *The Star* also pointed out in a leader column, this small triumph merely postponed the inevitable.

Labour tenants have also had to stand up to large companies. Forestry companies have moved into parts of South East Transvaal, evicting some labour tenants and employing others, insisting that those they retain must live in nuclear family units and also dispense with their large herds of cattle. Not surprisingly, such schemes have met with huge resistance – which in at least one case has paid off.

In 1987 the forestry company Hunt Leuchars and Hepburn, which had issued eviction notices to eight families, suddenly halted the process of eviction. HLH management met with the tenant farmers in a clearing in the forest and listened to the families explain the contracts which had allowed them to farm an area of land in return for providing labour, and how this system had operated for generations. At the end of the meeting the group personnel manager stood up and apologised to the families, admitting that the company had acted inappropriately.

He said every person present could stay on the farm for the rest of his or her life and the summonses would be withdrawn. He sat down. There was dead silence in the forest; this stretched on and on. He jumped up and said that perhaps they didn't believe him. He told them his name and his position and assured them that he had the authority to implement what he had said.

The villagers were too amazed to respond immediately. Then one after another they stood up and made tentative speeches of thanks. After he had left there was incredulous wild rejoicing.[19]

Makhosasane's goats

For Makhosasane Ndawonde eviction meant not only being dumped in a relocation area but a two-year nightmare, trying to recover her goats. She told AFRA her story:[20]

The government trucks came and took us to Nondweni. The official told us we should leave our livestock and collect them later. We were dumped in Nondweni and when I came back on the Saturday to collect my chickens and goats they had gone. I went to the farmer to ask him where my chickens and 300 goats were. He refused to tell me, so I went to the Commissioner in Greytown for advice. The Commissioner wrote a letter which I took to the farmer, but he refused to read it. I returned to the Commissioner, who advised me to take the letter to the Greytown police. The police took me to the farm but the farmer was not there. Next day I returned to the police, who told me they had spoken to the farmer's mother and she had told them the goats were at Muden; I should therefore go to the Muden police. I did this, but the Muden police said they knew nothing about the goats.

I returned to Greytown, where I met a friend who advised me that some goats had been seen on another farm near Inadi. I went there with the Greytown police and found 53 of my goats. The Greytown police said this was a case for the Muden police. I went with the Muden police to this same farm. The man in charge of the goats said I should pay R200 (£50) for my goats. I had no money and felt that the police should recover my goats. The police said they would find out about the R200 from the farmer where I had been staying and contact me.

They did not contact me, so I went back to the Muden police, who told me their van was broken so they couldn't collect my goats. They told me not to worry, because my goats were now under the protection of the law. I waited, but when I realised that the police were not going to contact me, I came to the Commissioner in Pietermaritzburg. This man advised me to go to the police as he said the farmer had no right to take my livestock.

I then went to the Greytown police, who advised me to see a Durban lawyer. I went to see one, who said I should pay R40 (£10) before he would do anything. I paid this money and he wrote a letter for me to take to the Muden police. The police said they would give it to the station commander. When I went back later, the police said that they had given the letter to him but did not know what had happened after this.

I returned to the lawyer and he told me he would solve the matter. I heard nothing from this lawyer and went to see him again in Durban, but I was told he was not there. I phoned twice the next day and was told he was not in. The second time the man who answered the phone told me I was a fool, so I decided not to return to this lawyer.

One day I was telling my story to a friend when a passerby heard it and advised me to go to the local KwaZulu MP. I went to him and he wrote a letter for me to take to the Commissioner. I went to the Commissioner's office and saw the magistrate, who advised me to go to the Muden charge office. I asked my father to do this as I was working. My father took this letter to the police and they said they would give it to the station commander.

I went back to the KwaZulu MP but he had nothing and he advised me to go back to the Muden police. My father went and the police told him that the station commander was on leave for a month. After a month my father returned to the police station but the station commander was not there. My father eventually saw the station commander, who told him to get a lawyer.

I went to see the KwaZulu MP, who took me to the Greytown magistrate. He advised us to go to the Greytown police station. We went there. The police said they had referred the matter to the Muden police. The station commander phoned the Muden station commander and asked me to come back the next day to find out what had occurred. The next day they told me that the prosecutor had decided the case was closed.

I went back to the KwaZulu MP and he said he would contact some lawyers and I should return to him after two weeks. I returned after two weeks and he said that I should come again in two weeks' time. When I went back he said that the lawyers said there could be no case because there was no evidence.

AFRA tried to help Makhosasane. The lawyer they approached on her behalf decided he could not take up the case for the following reason: notice might have been legally served on the complainant and her family; once the notice had expired and the magistrate had decided these people were illegally on the land, the family's stock was also illegally there and could be impounded by the farmer; once the complainant did not pay the R200 asked from her by the pound, they would have been sold by the pound and the money taken by the pound.

For Makhosasane it is clear whose interests the law serves.

5 THE STRUGGLE FOR BASIC RIGHTS

While other sections of the South African workforce have managed to win improvements in wages, working conditions, and legal protection over the last two decades, the vast majority of farm workers – along with domestic workers – remain out in the cold. The most basic rights – a living wage, healthy working conditions, security, reasonable accommodation, adequate education for their children, freedom of association – are denied them. The situation of most farm workers has in fact sharply deteriorated.

Obviously, the most effective way of challenging oppression on the farms is for farm workers to organise themselves into unions. But the struggle for union recognition is still in its infancy.

Organising farm workers

In August 1987, 1,200 tea pickers and other workers were dismissed from the Sapekoe Tea Estate in Richmond, Natal, for striking for better wages and conditions. Their principal crime, in the eyes of the management, was that they had taken their grievances to the Food and Allied Workers' Union (FAWU). Most workers on the estate earned R55 (£13.75) a month. Pickers worked ten back-breaking hours a day, six days a week, with minimal protection against dangerous agricultural chemicals. Other workers in the tea-processing plant suffered from exhaustion and chest ailments in the hot, dusty conditions of the factory. Accommodation was in single-sex hostels; in the women's hostel it was not unusual for four women to share a room with their young children. Although some women had rooms in their husbands' hostels, many couples were separated, and men visiting the women's hostels risked dismissal. One of the main grievances was the poor quality of the food, the daily rations consisting of rotten mealie meal in the morning and stew in the evening.

The women were particularly oppressed. As primus stoves were forbidden in the compounds, women had to walk long distances for firewood and rise early in the morning to heat water. Pregnant women were often evicted. A company nurse commented:

> They were told to give birth at home. This surprised me, because the woman was not pregnant when she started working for the company. Why is it that now because she is pregnant she has to be evicted? ... Some women are taken back when their children are ten months old and can feed themselves. It is difficult for a pregnant woman to carry a basket. At other places, other than Sapekoe, a pregnant woman is given a light job to do in the yard.[21]

The company nurse was one of 12 shop stewards elected (the others were all men) to take the workers' case to the management after the workforce had joined FAWU. Stonewalled by the management, the representatives advised strike action. An increased bonus of one cent per kilogram for picking was the management's only response and when this was rejected, police and soldiers moved in to evict the workforce.

> They assaulted us, took us out of the gates and told us to come back and get our belongings and our wages and then get lost. They dumped other people at Richmond station. Of course, we lost lots of belongings in the chaos. Also, some of our pay packets were R20 (£5) or R30 (£7.50) short, because they knew we couldn't complain.[22]

The disastrous outcome of unionisation and strike action at Sapekoe is typical of most attempts by South African farm workers to secure a better life with union backing. In a few notable cases, however, political and economic circumstances have been in the workers' favour. Among the few victories was the 20-day strike during May 1989 at Zebediela Citrus Estate in Northern Transvaal.

Workers at Zebediela, which is known as the world's largest citrus farm, were required to pick 56 large bags – about R450 (£112.50) worth – of oranges a day. For this they were paid R5 (£1.25). They were allowed just 12 days' annual leave, unpaid. When, at the beginning of the 1989 season, the management issued pickers with larger bags to fill, the Zebediela workers downed their baskets, calling for a return to the old-size bags, a 45 per cent wage rise, and recognition of their union, the National Union of Farmworkers (NUF). The management responded by dismissing the whole work force and bringing in the police, who unleashed their dogs on the workers; fifteen people were hospitalised with wounds from dog bites.

Workers were ordered to evacuate their company houses within six days. For the children, eviction meant not only homelessness but total disruption of their education, since they all attended school on the estate.

The NUF went to the Black Lawyers' Association in Johannesburg, who were able to turn the tide by seeking a ruling by the Supreme Court against the dismissals and eviction. The court application was postponed when the management, finally persuaded that the workers were highly organised and united, agreed to meet their representatives. Negotiations lasted three days, eventually ending in an agreement by management to reinstate dismissed workers, to pay a monthly wage increase of R30 (£7.50) across the board, and to negotiate a contract of service and disciplinary and grievance procedures with the union's lawyers.

Although the wage increase was a paltry one by comparison with settlements in other sectors, for the lowest paid workers at Zebediela it actually doubled their monthly pay. And the greatest victory was in compelling such a large and powerful employer to recognise that it could no longer treat its workforce with contempt and get away with it.

Another success story comes from the Verlegen Estate, belonging to Anglo-American Farms, in the Western Cape. There, FAWU was able to use the threat of disrupting an international wine auction, planned for January 1990. Anglo-American had refused to meet FAWU throughout 1989, but the threat of a mass picket and international media coverage led to sudden recognition of the union and a promise to review the conditions under which casual workers were employed.

Unions organising farm workers

Several unions are attempting to organise farm workers. The Food and Allied Workers Union (FAWU), affiliated to the Congress of South African Trade Unions (COSATU), launched the Farm Workers Project in 1985 and claims 2,200 members on the farms. The National Union of Farmworkers (NUF), affiliated to the National Council of Trade Unions (NACTU), was also formed in 1985, and claims 20,000 members.

Other unions include the Paper Print and Allied Workers' Union (PPAWU), a COSATU affiliate with 5,000 members in forestry plantations; the Orange Vaal General Workers Union (OVGWU), an independent union claiming 4,000 members on farms in the 'maize triangle'; the National Union of Farm and Agricultural Workers of South Africa, launched in 1988; and the United Workers Union of South Africa (UWUSA) which is aligned with Inkatha, the Zulu nationalist political movement led by Chief Buthelezi, Chief Minister of KwaZulu homeland.

Congress of South African Trade Unions (COSATU), 1987. (Paul Weinberg/Afrapix)

The number of unions organising farm workers is in sharp contrast to their organisational gains; indeed, the large number of different unions in the sector is one factor which weakens their effectiveness. Most successes have been on large farms run by multinational or state-owned companies, and even these have been few and far between.

There are numerous reasons why unionisation is difficult. Although large corporations have moved into agriculture in recent years, most of the farms are still individually owned. Whether obviously oppressive or more benevolent in their treatment of workers, almost all private farmers are opposed to their workers being unionised. The Labour Relations Act, which grants freedom of association to other sectors of the South African workforce, excludes farm workers (and domestic workers), and farmers are not obliged to negotiate in the case of a dispute. Most farmers are deeply suspicious of 'outside interference' in the affairs of 'their workers', and refuse to allow union officials on their land; union organisers who set foot on the farms can be served with trespass orders and, at worst, risk life and limb. The workers themselves are in the most vulnerable position. Those who attempt to join unions frequently suffer severe harassment or are summarily dismissed.

In 1989 the farm owners' South African Agricultural Union (SAAU) brought out a booklet outlining its stance on unions. It states that unions have a legal right to act on behalf of farm workers, and may take legal action based on workers' common law rights. However, it emphasises that although employers would have to talk to unions in these circumstances, they are not obliged to negotiate with them, and advises strongly against recognition agreements, as these would give farm workers 'rights and powers via common law which, in terms of the Labour Relations Act, were not intended for them'.

COSATU's decision in 1985 to start a Farm Workers Project under the umbrella of the Food and Allied Workers Union, rather than a separate union for farm workers, was based on the need to give farm workers maximum support through links with other workers. The principle was that food industry workers, who are protected by the Labour Relations Act, would be able to take solidarity action if farm workers were victimised for trying to form unions and demand changes.

The Farm Workers Project (FWP) has concentrated on organising workers on estates and factory farms owned by multinational corporations. These have industrial relations policies which allow unionisation, and also own industrial enterprises where FAWU's well established membership can exert pressure through sympathy strikes.

Some agribusinesses and state farms and even some family farms are legally defined as factories rather than farms. This is the case when the main business is a processing operation, as on poultry farms where the fowl are not only bred but also slaughtered, cleaned, packaged and frozen. These farms are therefore covered by the Labour Relations Act and are easier targets for unionisation. The FWP has made some inroads into such farms. In the Western Cape, for example, about twenty Farmfare and Nulaid farms have recognised FAWU. However, this does not represent large numbers of unionised workers, as there are only 20-30 employees on each of these farms. On other chicken farms in the same area, such as Rainbow and Country Fair, the majority of workers have joined the FWP, but there are no recognition agreements with the management.

The FWP has not tried to recruit members on the majority of private farms, believing that it is pointless to do so under the present feudal conditions and without changes in labour legislation.

Johann Hamman, a worker at the Stellenbosch Advice Office, was an organiser with the FWP until February 1989 and had quite a few successes in larger fruit and wine farms in the Western Cape. He gave the following examples of successful action.

A woman of 55 with two children who was evicted when her husband died of a heart attack was reinstated after workers went on a march. On another occasion workers went on strike to demand R40 (£10) a week (the going rate was R26 (£6.50)), window panes in the houses, and less swearing at workers. The strike happened at the right time of year – pruning time – and farmers agreed to the demands as they could not find other workers. Another strike was organised when 30-40 people were dismissed from a vegetable farm. They picketed outside a restaurant and again won their case.

One of the FWP's major problems as a young organisation with a small membership is lack of funds. Workers need training in negotiating techniques, as well as general education about trade unions. It is important for them to meet with other farm workers in the area to share ideas and problems, and gain from a sense of solidarity. But membership subscriptions are not sufficient to meet the cost of laying on courses and hiring transport to bring workers from the farms. COSATU was able to fund the FWP only for the first year.

The other main problem, according to Johann Hamman, is the urban bias of FAWU. At present there is an organiser in Natal but none in the Orange Free State, Transvaal or Cape. The FWP now has to rely on the committees set up on certain farms.

The other major black labour federation, the National Council of Trade Unions (NACTU), decided, unlike COSATU, to form a union specifically for farm workers. The National Union of Farmworkers (NUF) has taken a different approach to organisation from the one adopted by the FWP, accepting individual members from private farms and attempting to take up their complaints with employers. It has secured only a small number of recognition agreements.

Philip Masia, organiser for the Orange Vaal General Workers Union, explained how the union took up the cause of farm workers on a large farm near Uitriucht owned by Anglo-American. He told an Oxfam fieldworker, 'When the workers on the Anglo-American farm approached me to ask if they could join the union, I told them no ... but they could *form* the union.' In 1984, after many struggles, Anglo introduced contracts for the farm workers, but these were negotiated with individual workers. The unionised farm workers managed to gain certain concessions: they won the inclusion of an agreement on hours of work and overtime, and threw out two clauses, one requiring workers to pay for broken machinery and another requiring any child over 16 not working on the farm to leave.

The unionisation of the Anglo-American farm encouraged workers from private farms to apply for membership. Philip Masia echoes other organisers in emphasising the difficulties of organising on these farms. The union can help only when there are dismissals or unfair treatment. It has successfully intervened in assault cases. These are common in the Vaal triangle at the end of the maize-growing season when farmers try to make workers run away before they are paid.

The United Workers Union of South Africa (UWUSA) has been described by agricultural researcher, David Cooper, as a 'sweetheart union', affiliated as it is to the KwaZulu political organisation, Inkatha, and having a reputation for collaboration with employers, as well as violent attacks on members of unions affiliated to COSATU. After the mass dismissals at Sapekoe, it was UWUSA that the management called in to recruit a new workforce.

Access to other outsiders

Your Guide to Farm Workers and the Law Book 2 provides a list of nearly 200 local advice offices, legal clinics and other organisations which offer support and help to farm workers. A number of these specialise in rural affairs, and are often in a better position than trade unions to help meet the individual needs of people working on privately owned farms. They take up a wide range of cases, including pension and compensation claims, and challenge such abuses as assaults, extortionate cattle impoundings, and other illegal conduct by farmers. Although they are particularly concerned with the plight of individuals, their families and communities, they are also able, by ensuring that cases go to court wherever possible, and by engaging in lobbying and media campaigns, to expose the harsh conditions suffered by farm workers and broadcast their struggles and successful actions.

AFRA and the Weenen Community Project

The Association for Rural Advancement (AFRA) works with evicted farm workers in the Mngwenya Valley, south of the town of Weenen in Natal. Its work is an example of appropriate action by an 'outside' organisation that recognises and supports farm workers' own self-help initiatives.

The Weenen area is dry thornveld country, where labour tenancy was the dominant form of farm labour until it was banned from the area in 1969. Mass evictions followed, but a modified form of labour tenancy, even more exploitative than before, has continued. However, many of the families who survived the state-sponsored removals of the early 1970s have since been thrown off the land by individual farmers.

A 'temporary' camp for evicted farm tenants in Weenen, Natal, November 1988. (Gill de Vlieg/Afrapix)

The tenants who were left formed the Weenen Community Project. It aims to raise the standard of living of farm dwellers, encourage negotiations between farmers and employees and improve relationships between them, resist evictions and discourage the drift of rural workers to the towns, support self-help projects, organise legal aid, and provide training for handicraft and home industries.

The Weenen farm workers asked AFRA to help them in their struggle to resist eviction. In 1988-89 AFRA's legal sub-committee managed to ensure the postponement of a number of prosecutions for 'squatting'. It also lobbied the Natal Provincial Administration to review its plan to move evicted families to the Weenen Emergency Camp (a 'temporary', 'emergency' camp set up twenty years ago to resettle urban dwellers from Weenen). At the same time, it ran a sustained publicity campaign against the plan in the local press.

With the cooperation of the Weenen farm workers' committee and the local Catholic church, AFRA established an advice centre in the town of Weenen to provide more systematic para-legal support in dealing with evictions, cattle impounding, pensions, Unemployment Insurance Fund payments, and other social problems.

Sealed off from the outside world

Organisations such as AFRA can usually help only if farm workers are able to come to them with their problems. Where workers live on the farms, it is often extremely difficult for outsiders to gain access to them in their homes. Most farmers insist that without their permission outsiders have no right to visit workers, and many are prompt to serve trespass orders if this rule is infringed.

The attempt by farmers to seal off their workers from the outside world is particularly inhuman when workers have suffered personal loss or injury, and are denied the right to receive comfort and condolences from sympathisers. One such case was described in a letter to Oxfam recently by Dorothea Russell, wife of the Anglican bishop of Grahamstown. Oxfam field workers had accompanied Mrs Russell on a visit to the M. family, who work on a farm belonging to Mr K. Some weeks earlier, Mr and Mrs M's only son had been killed in an accident by another farmer, Mr R., whose truck ran into the boy on the road. This farmer offered no apology and no help to the family in their bereavement, and had not even been to see them.

After taking the Oxfam visitors to see the family, Mrs Russell received an irate phone call from Mr K. In threatening tones, he said that he understood from his informants that she had brought a vehicle on to his land without his permission and that foreigners had accompanied her. The farmer was particularly incensed that the 'foreigners' had taken photographs; it was acceptable to photograph monuments and buildings, but not to take photographs of people without his permission. He made threatening noises about farmers having their own 'civil defence', and said they were on the alert. Dorothea Russell commented:

> I was so outraged at the implication, and asked him, 'What on earth did you think I was going to do? Did you think I wanted to burn down your farm? Or rob you? Or commit murder? (...) What was *wrong* with visiting that poor family? We did *not* keep them back from working for you.' They were just there – at home. And they were overjoyed at our visit.
>
> Yes, I knew it, in theory, but now I was experiencing in practice that a farmer regards the men and women working on his farm as his *property* – one must get his permission to visit them outside their working hours. Well!! Is that not prison conditions? I told him I didn't want to hear any more about trespass and 'the law'. That was the Ms' home – grandmother's, aunts', uncles', parents' and children's extended family home. And one has to get his *permission* to speak to them at home while they mourn the loss of Zola, their only boy ...?!

The role of the churches

Although church workers do sometimes meet hostility from farmers, in general they have greater access to farm workers than any other 'outsider' normally has. In also having access to the farmer – since most white farmers claim to be Christian and are regular church goers – they are in a unique position to hold dialogues with both parties. The South African Council of Churches (SACC), after passing a resolution in 1989 expressing solidarity with farm workers, outlined a practical programme of action. This urged member churches to use their special relationship with farmers and workers in order to tackle such issues as violence against farm workers, to help prepare the way for unions to organise the workforce, to inform workers about their rights, and to campaign for facilities to reduce illiteracy.

Some of the churches own large tracts of agricultural land. While some church people are extremely active in their efforts to help farm workers, some church farms are themselves guilty of exploiting workers. An SACC field worker, Thekiso Tlhacoane, is acutely aware of this contradiction: 'Often there is no difference in the treatment of workers who are working on church-owned land and those who are on the farms owned by large corporations.'

Southwell Mission

The Anglican Church holds the title deeds for a piece of land 30 kilometres south of Grahamstown, in the Eastern Cape. On this land stands a tiny church with a leaking roof, a three-roomed schoolhouse serving over 200 children from 29 farms, a row of ramshackle houses for teachers, and a communal kitchen. The classrooms are dark and ill-equipped. There is no electricity and there are no toilets. The children go out to the veld – a class at a time. Close by is a spacious school for 15 white children, complete with electricity, tennis court, and audio-visual equipment.

Dorothea Russell has started working with the community, helping people to improve the quality of their lives. The following reports trace the struggle to win farm workers' confidence and pursue the project in the face of considerable opposition:

> First we applied to the Department of Education and Training (DET), which offers funds for the building of classrooms. After lengthy consultations over the design of the new schoolrooms, after acquiring a builder and a clerk of works and after many other meetings, we were dismissed with nothing, on the grounds that 'the school now stands on Church land'. Since the inspector, a DET employee, knew about the Church ownership from the beginning, this amounts to calculated deceit on the part of the DET.
>
> We then considered repairing the dilapidated church. The essential needs were a roof, a ceiling and a small vestry. The utterly poor congregation was deeply pleased at the very idea of care and concern, but there were grave reservations. The local African priest advised us to find Mr Ntsikizeni Buni (a farm worker from a nearby farm, now living in Grahamstown) and to listen to his story ...
>
> Eventually, through this person and that, Mr Buni was tracked down in a backyard wattle and daub structure, and he told us his experience. In March 1988, the farmer who had employed Mr Buni for at least 30 years suddenly changed in his attitude towards him. He began making taunts and jibes about Archbishop Tutu. 'He's only a pagan. He's not even a Christian.' Mr Buni remained loyal to his Archbishop, insisting that he was a Christian.
>
> The next month, over a mere trifle (the chicory should have been weeded in one way and not in another) the farm worker was blasted with: 'Get out! Go to your bl... f...ing church! Go to your Archbishop! Or go to Russell in Grahamstown!' With that, Mr Buni, his wife, their

five daughters, two grandchildren and Mrs Buni's elderly mother were dismissed from the farm which had been their home for two generations.

A type of fiefdom exists on these farms. A complaint or a quarrel over a minor issue can have dire consequences. Even disapproval of the Church to which you belong can be fatal. One priest from Grahamstown who wanted to minister to farm workers and asked the farmer's permission to hold a Sunday service was told, 'If you set foot on my land you will be shot.'

You can see why we had to abandon the project to improve the workers' humble church. Finally we decided that rebuilding the communal kitchen was one need that could be met, with some assistance, but would not be threatening to the farm workers. The farm workers have decided to provide the labour. Mr Buni, the evicted worker, will work full-time on the job, with others helping part-time.

The Secretary of the farm workers' community wrote to Oxfam:

Now that we have begun, we want to move on and progress with what we need the most. We want to take our children out of the church building and put them in a schoolroom. We want to use our church for worship. We need toilets for our children. There are no toilets at the school.

We keep pigs and chickens. If we had a proper place to keep the animals, we could teach our boys and girls how to rear healthy livestock. But we need training. We need instruction in building ...

Farm workers and our families are cut off from all places of training, and hope for our children's future. Our young boys would like to be trained in agriculture, machines and carpentry. The young girls want to be trained in useful skills.

Dear friends of ours, walk together with us in the darkness. Sala Kakuhle! (Stay well!)

It is encouraging to note that the farm workers of Southwell are now linked with the new cooperative at St John's Hillside (described in Chapter 4) – a move intended to help both communities to overcome their sense of isolation.

The farmers' response

While feudal conditions persist on the majority of farms, the combined pressure of unions, support organisations, sections of the church, media publicity, successful court actions and international boycotts has caused some farmers to improve conditions for their workers. Some of these farmers undoubtedly recognise the need for change, and are concerned about the welfare of their workforce, but probably the strongest motivation is the desire to head off the unions and other 'trouble makers'.

Popular farming journals devote considerable space to reports of farmers' efforts to improve conditions. The following, from *Farmers' Weekly* (9 May 1989) is characteristically paternalistic:

> Fifteen Bothaville farmers have achieved remarkable success in improving the quality of life of their farm workers within a period of five years. The Bothaville North Community Development Association which has been active on 15 farms spent R120,000 (£30,000) on housing for workers on five farms, with the result that 86 per cent of the families are living in brick houses.
>
> A primary school situated within easy reach from all farms in the area has been enjoying 100 per cent attendance. Another one to cater for up to standard 7 is envisaged.
>
> Close co-operation with supporting health services has limited the incidence of disease, child deaths and unplanned pregnancies to a large degree. Most farm workers have completed one formal course at either Boskop or Vetsak Training Centre. In co-operation with the Department of Manpower, 120 wives of farm workers have been trained in needlework and knitting. Much is also being done to promote the spiritual well-being of farm workers, and lay preachers are specifically trained for this task during training sessions.

Although there are obviously important gains here, such as improved health care and educational provisions, there is no mention of wages or worker organisation.

The South African Agricultural Union (SAAU), the state, and the farm owners' Rural Foundation are actively devising strategies to head off unionisation and improve relations between farmer and farm worker, thereby protecting their profits. The SAAU has formulated a code of conduct which it is encouraging farmers to observe. *Farmers' Weekly* reported in May 1989:

> This is aimed at uplifting the working conditions of farm workers generally, and easing the strain on farmers. According to SAAU's deputy director of general services, Kobus Kleynhans, such a document is likely to be far more effective in reaching farm workers' goals than could possibly be achieved through a trade union or other movement.

Supported by the state and by the SAAU, the Rural Foundation is an organisation of farmers committed to upgrading conditions on the farms. It was set up at the end of 1982, by the government and a large farmers' co-operative, as an 'apolitical and non-racial organisation'. It began working in the Western Cape and is now also operating in Western Transvaal. Most of the farms it works on are labour-intensive, with a workforce of 50-100 or more, and export-oriented – and therefore vulnerable to the effect of international sanctions. In 1988, it was involved in projects on 2,790 farms (a 32 per cent increase on 1987) with 35,917 farm dwellers. Out of a total number of farms throughout the Republic of approximately 58,000, this represents less than five per cent. But where it has gained access, the changes have been quite considerable.

On some farms modern housing, schools, creches, and sports and leisure facilities have sprung up, where none existed before. There is reportedly a corresponding shift in attitudes among farmers, signifying the end of the old master/servant relationship. How far this goes is debatable. As one farmer put it,

'We no longer talk of having so many kaffirs on our farms. We talk of our personnel. Farmers no longer simply tell their workers what to do. We meet each morning to discuss the day's work.'

The Rural Foundation works closely with the Department of Health's Population Development Programme through rural community development associations, such as the one at Bothaville, cited by *Farmers' Weekly*. It works on farms by invitation only. If a farmer decides to take up its programme of reforms, he foots about 25 per cent of the bill himself, the lion's share being covered by the Department of Health, supplemented by business sponsorship. Among those corporations throwing their weight behind the foundation are Mercedes SA and Volkswagen SA.

Herman Baily, of the Rural Foundation, is quite straightforward about the reasons for the programme: 'I believe it makes sense to invest in people, especially in this labour-intensive branch of agriculture. This kind of investment can amount to 37 per cent of farmers' outgoings. So it's important that the investment is well managed.' He describes the organisation's aims as, firstly, to encourage farmers' participation in 'the development of communities'; secondly, to improve the working conditions of farm labourers; and, thirdly, to improve their social and living conditions. Arguably, it is this third aim that has received most attention so far.

One aspect of the Rural Foundation's approach to reform, sharply criticised by unions and democratic organisations, is the setting up of 'liaison committees'. These are groups of workers, selected by management to act as go-betweens for both workers and management. They are seen as the farmers' alternative to genuine workers' representatives – trade union shop stewards – and as a means of controlling the workforce more conveniently than by compulsion, which demands costly supervision.

'We're doing all this because of sanctions'

Ben Schoeman, a sociologist conducting research into the Rural Foundation, gives his views on the farmers' reasons for carrying out reforms and on the limits to such reforms:

> The important question here is sanctions. Farmers have frequently said to me, 'We're only doing all this because of sanctions.' There are regular visits from overseas business people who come and look at the situation here. And farmers are also told by factory workers, 'We won't pack your fruit any longer, unless you improve conditions.' I believe farmers are motivated by fear, and that because of sanctions some of them have had real difficulty in selling their products ...
>
> The so-called liaison committees are mainly concerned with social problems, such as housing, rather than working conditions, so they're not much help to farm workers. On some farms they have become an instrument of control for the farmers. As long ago as the 1970s they were setting up liaison committees in order to organise the workers before the unions arrived.

The workers have gained, in that they need not now be so fearful if they have something they want to say to the boss or the manager. I've noticed that on all the farms I've visited. On the other hand, this new friendliness is a change of strategy. 'Just come to me if you have a problem and we can talk about it.' That means you're not likely to oppose the farmer.

In the past farm workers had absolutely nothing, and when people have nothing at all they don't fight either. What's happening now through the Rural Foundation is very much to the workers' advantage. No one can say it's a bad thing. I believe that, for example, COSATU's Farm Workers Project will gradually become stronger on the Rural Foundation farms, as workers will soon become frustrated. At the moment they have liaison committees addressing social conditions, but not wages, for example. On each farm that I visited, I asked, 'Are your wages sufficient?' and they always said, 'The money has never been enough.'

When I visit farms I ask the farmers a direct question: 'Don't you think that the fact you are giving workers more responsibility and control will have a snowball effect and they will one day take over the whole farm?' The farmers say no, that the changes taking place are within certain limits and these limits are set by the farmers. But I think that even if workers don't see through that now, they will perhaps in two years' time. Workers aren't stupid, they're even extremely clever, and they very quickly know what's wrong. They know now that they're being given fine new houses and sports grounds. But one day they will also know that more belongs to them. Even today it's very clear to them how far the liaison committees can go.

I think the Rural Foundation is a midwife in a process it really wants to prevent.[23]

The Winelands Commitment

Another farmers' initiative is the Winelands Commitment, a programme of rights for black and coloured workers, promoted by a handful of the country's leading wine producers. Signatories to the Winelands Commitment pledge themselves to various principles which not only grant workers' rights within the current political context, but also call for the end of apartheid practices. So, along with the right to freedom of association and a guaranteed wage conforming to 'acceptable norms', it concedes the right of blacks and coloured people to own agricultural land.

At the end of 1989 there were only four signatories to the commitment, but the group is an articulate one which may steadily gain influence. They have agreed on a minimum wage of at least R500 (£125) a month, and have a policy of unilaterally applying the Basic Conditions of Employment Act, although it excludes farm workers. Their progressive policies are not well viewed by the Rural Foundation, whose co-founder, the KWV co-operative, is the country's largest wine producer.

Farm workers organising for themselves

Such concessions are valuable, but in the end, it is more important that rural communities are able to set their own agendas for change. The challenge for both South African organisations and external development agencies such as Oxfam is how best to support what is often a long and slow process of organisation in these communities. But it is beginning to happen. One example is in the Eastern Transvaal, where for several years Oxfam has supported the Sizabantu Action Committee, which gives para-legal advice to communities in a large area around the township of Amsterdam. A legal clinic was also set up by the Legal Resources Centre in Driefontein. Increasingly they have taken up the problems of farm workers, labour tenants, and villagers living in the forests. (This work, concerned with evictions, assaults, and the need for schools, has not been without its risks. Jotham Zwane, the chair of Sizabantu, was threatened with eviction for being 'idle and undesirable'. Shortly after the local magistrate had found that Zwane could not be classified as idle because he ran his own transport business, the trucks were set on fire and destroyed. Zwane was duly found to be idle, and evicted. Later he was detained for five months under the emergency regulations.)

Nineteen rural communities in the Eastern Transvaal have come together, through their contact with the legal clinics, to form a committee to tackle their common problems. Oxfam field workers attending the Eastern Transvaal Network's first two-day workshop, facilitated by TRAC, in February 1990, reported:

> Although they met at the end of a hard week's work and long journeys to the venue, the participants' enthusiasm kept them working till nearly midnight. They were up again at six to make sure they got through the agenda. Representatives, some from rural townships, some from farms and some from so-called black spots, identified a long list of common problems, foremost of which is their isolation. Future workshops could play an important role in overcoming this. They plan to invite people who will help with administrative and organisational training. Spirits were high as people packed into trucks for the long journey home. We felt privileged to have been part of something special.

The first meeting of the Eastern Transvaal Rural Network, February 1990. (Rona Alexander/Oxfam)

6 LOOKING TO THE FUTURE

What can be done to tackle the poverty and oppression of farm workers' lives? Short-term programmes are urgently needed to address the worst abuses of the system and ensure minimal protection and rights for farm workers. Beyond this, more fundamental change is called for in a post-apartheid, democratic South Africa, and strategies to that end must inform the negotiations currently taking place between government and representatives of the mass democratic movement.

In the process of reform, what are the most effective contributions that can be made by unions, service organisations, churches and – most importantly – farm workers themselves?

Protective legislation

In its submission to the National Manpower Commission on farm labour in 1982, the Farm Labour Project made a wide range of recommendations on behalf of farm workers. Among the most important demands were the following:

- enforceable written contracts;
- wages competitive with other sectors, and the extension of the Wage Act to include agriculture, with special provision for ensuring adequate rations and other basic material needs;
- a basic 46-hour week, payment for overtime, and adequate rest breaks;
- holiday pay and leave benefits in line with the Basic Conditions of Employment Act;
- social security and welfare grants comparable with those of workers in other sectors, and the extension of the Unemployment Insurance Act to include farm workers;
- the extension of health and safety regulations;
- compulsory education for workers' children up to Standard 6, and enforceable restrictions on child labour;
- the right to free association and collective bargaining, guaranteed by extending the Labour Relations Act to include farm workers.

By June 1990 the findings of the National Manpower Commission had still not been made public, and no changes in labour laws had occurred. However, a year earlier the Director of Manpower had announced that the government intended to introduce new legislation, and in March 1990 a member of the NMC's legislative committee confirmed to the media that the process of hammering out a new statute to replace the Labour Relations Act was underway.

If enacted, such a statute would at least be a crucial first step in the campaign for farm workers' rights.

Strategies for the unions[24]

Unionising farm workers has so far proved extremely difficult. Andrew Ball, a researcher with the Farmworkers Research and Resource Project, argues that the unions have failed to develop careful enough strategies for organising workers: 'Success is only likely if unions develop clear strategies based on a sound analysis of the different sectors, crops and markets, and an analysis of the agribusinesses that are increasingly dominating many sectors.'

He points out that the pressure of international sanctions (real or threatened) on foreign-owned multinational companies can help unions make headway with organisation. Among these companies are some chemical and pharmaceutical companies which hold a small amount of agricultural land where they test their products. These could become a prime target, for organisation as well as solidarity work, as the pesticides and herbicides they produce are responsible for a shockingly high number of injuries and deaths on the farms.

Organising workers in South African agribusinesses, which tend to have a variety of holdings in order to dominate a particular sub-sector, requires concerted efforts within all the linked operations. Since many private farms are contracted to supply a single monopoly company, organised workers in factories or large estates could support the struggles on private farms.

State-owned research farms at various universities and colleges could also be targeted, as the government and sectors of the South African Agricultural Union are backing the 'liberal' Rural Foundation and are unlikely to want repressive tactics used against workers on university farms which are supposed to be promoting good labour relations.

If workers are to be able in time to become independent farmers, unions organising farm workers must redouble their efforts in the area of education and training, from basic literacy and numeracy to sophisticated management techniques.

A challenge to the churches

Although not major landowners, the churches own small amounts of agricultural land throughout the country. Many church farms currently leased to white farmers could be put under worker control; a few schemes of this kind have already been introduced in recent years. Pilot schemes are a very useful way of exploring problems and potential in preparation for the increase of black-run farms that liberation will bring. After long years of working for white farmers, black workers may find difficulty in taking control, and may need support in their efforts to learn new responsibilities. The churches may also have a role to play here, in consciousness-raising, training, and facilitating.

The churches could also increase their liaison and lobbying roles, and help to move the needs of farm workers higher up the political agenda.

Land reform and agrarian reform[25]

The Freedom Charter's assertion that *The land shall be owned by the people who work it* represents a formidable challenge to any post-apartheid government. Yet, for South Africa's rural poor – farm labourers and their families, migrant workers from the bantustans, the dispossessed and the landless – justice will require more than merely eliminating flagrant abuses. It also demands fundamental changes both in land ownership and in production relations on the land.

As majority rule draws nearer, the complexities and constraints affecting any attempted rural reform are emerging more clearly. Policy debates are taking place on the often contending claims of political and economic stability, and land redistribution and agrarian reform. Maintaining production will be the major agricultural priority for a new government, and any land reforms will have to be made with this need in mind.

The starting point for land redistribution could be to acquire unproductive white farms owned by absentee landlords. If, as in Zimbabwe's Lancaster House agreement, land can be appropriated only on a 'willing-buyer willing-seller' basis, underutilised or abandoned farms could be an obvious first target, partly because these will be the cheapest to buy.

Critics point to the dangers and shortcomings of this option. Farms abandoned by white farmers are not uninhabited; they are usually heavily populated by farm worker families. Underutilised or abandoned land is usually ecologically marginal, and even heavily-subsidised white farmers have failed to make it productive. As the sociologist Tessa Marcus argues: 'The state would have to divert large amounts of limited resources into these areas with very little gain, while the richest land areas would remain untouched, firmly in the hands of private, mostly white, high-capitalised farmers.'

State intervention in land and agrarian reform, which remains a central tenet of the policy of the African National Congress (ANC), would involve making inroads into the fertile, productive core of white South African agriculture. Buying such land at market prices could prove prohibitive, and goes against the grain morally. Many progressive economists feel that it is unacceptable for the repayment of former land owners to become a national burden on the labouring population, whose exploitation has created the farmers' wealth.

One proposal being debated among ANC economists is a system of revaluing land on the basis of both productive capacity and profitability, and for compensation for nationalised land to be paid in the form of government bonds. The ANC is committed to a flexible, non-dogmatic approach. Cooperative farming, state farms and leasing back are all being considered, depending on suitable circumstances. Whatever methods are chosen, the state will need to provide a range of assistance to help black farmers establish themselves.

Aninka Claassens, formerly a worker for the Transvaal Rural Action Committee, cautions that land reform will work only if it responds to rural people's own conception of land rights: 'What I've seen in rural land struggles is that people fight for their notion of land rights, a notion which is much stronger than anything this present government has managed to impose.'

She proposes a radical redefinition of land rights that challenges the notion that private ownership through title deeds constitutes the only legitimate claim to land. She feels that the new government could play an important role as mediator when there are rival claimants to the land.

> What I'm basically talking about is a situation where land is seen in its historical context and the state develops a set of parameters to mediate people's claims to land – going right back to all the different groups that have lived on the land and balancing people's claims in terms of birthright, in terms of occupancy of the land, in terms of their productive usage of the land. The system would acknowledge the right of inheritance.

State mediation of this kind would be quite different from prescriptive, centralised planning such as Tanzania's *ujamaa* system and South Africa's own 'betterment' schemes, which Aninka Claassens argues have been disastrous. The essential point is to adopt a diversified approach.

> There are enough different kinds of struggles to indicate the way forward. For example there are enough struggles of, say, tenants (on white farms) and bantustan peasants for the extension of family farming to show that that's a major demand and a productive course. There are other indications from people in more mechanised farming areas that they see the future resolution of the land problems in their areas to do with collective farming. And there are other instances, particularly in church land projects, where farm workers have not had the kind of confidence that they want to run the farms themselves. They see the issues as better wages, more security and better working conditions.

Philip Masia of the Orange Vaal General Workers Union puts farm workers' demands in simple terms: 'We don't need to teach our members the Freedom Charter. They have a passion for land. What they want is land to till and animals to keep ... It's a modest type of demand.'

The last word should go to farm workers and rural dwellers. For some, like 63-year-old Mr Zungu from Weenen, all that is asked for is a place to die in peace:

> Now that my back is broken like this, that I am so old, I am sick, I am of no use to the farmer any more – I must go! Where will I go to, as sick as I am? How am I, an old sick man like me, to start all over a new life again? Why doesn't he let me die here on the farm?[26]

For others, like Petros Nkosi from a rural community in South Eastern Transvaal, the struggle is for the land:

Our purpose is the land, that is what we must achieve. The land is our whole lives, we plough it for food, we build our homes from the soil, we live on it and we are buried in it. When the whites took our land away from us, we lost the dignity of our lives, we could no longer feed our children. We were forced to become servants, we are treated like animals. Our people have many problems, we are beaten and killed by the farmers, the wages we earn are too little to buy even a bag of mealie-meal. We must stand together and help each other and face the Boers. But in everything we do we must remember that there is only one aim and one solution and that is the land, the soil, our world.

What does the future hold for these farm workers' children from Kei Road, Eastern Cape? (Gill de Vlieg/Afrapix)

NOTES

1 *Daily Dispatch*, 27 April 1989.

2 Reprinted with permission from *Native Life in South Africa* by Sol Plaatje (Harlow, UK: Longman, 1987, pp. 57-59, abridged).

3 Details of this historical fraud are given in *Race, Propaganda and South Africa*, by John C. Lawrence (London: Gollancz, 1979).

4 This section draws extensively on *Modernizing Super-exploitation: Restructuring South African Agriculture* by Tessa Marcus (London: Zed Press, 1989).

5 *Forced Removals in South Africa*, the Surplus People Project Report (Cape Town: SPP, 1983).

6 This section is based on *Working the Land: A Review of Agriculture in South Africa* by David Cooper (Johannesburg: Environmental and Development Agency, 1988; chapter 3).

7 Quoted in Cooper, op. cit.

8 This section is based on 'Four industrial laws' in *Your Guide to Farm Workers and the Law, Book 1* compiled by Annemarie Hendrikz (Cape Town: Rural Legal Services Project, 1988).

9 'Abandoned Babies? or In Search of Victoria', by N. Mpande, C. Nolutshungu, and J. Davies (Diocese of Port Elizabeth, November 1988).

10 Reproduced with permission from *Schattendasein: Farmarbeiter in Sudafrika (Living in the Shadow — Farmworkers in South Africa)*, by Bettina von Clausewitz (Hamburg: Evangelische Missionwerk, 1990).

11 'Abandoned Babies', op. cit.

12 Worker Education Project/Taung Migrant Project, report for November 1989-March 1990.

13 Adapted and abridged from 'There you are under the cows' by Anna Mazibuko, in *Vukani Makhosikazi: South African Women Speak* by Jane Barrett *et al.* (London: Catholic Institute for International Relations, 1984, pp. 69-76). Reprinted with permission.

14 *Child Labour in South Africa* (London: Anti-Slavery Society, 1983).

15 *Modernizing Super-exploitation: Restructuring South African Agriculture*, op. cit.

16 Abridged from 'Working the land is not child's play' by Pat Devereaux, *The Star* (Cape Town, 24.6.1989).

17 This section draws on *Umhlaba: Rural Land Struggles in the Transvaal in the '80s*, by Aninka Claassens (Johannesburg: Transvaal Rural Action Committee/ Black Sash, 1989) and *A Toehold on the Land* (Johannesburg: Transvaal Rural Action Committee, 1988).

18 Quoted in *A Toehold on the Land*, op. cit.

19 *Umhlaba: Rural Land Struggles in the Transvaal in the '80s*, op. cit.

20 Association for Rural Advancement, Pietermaritzburg, report no. 21, July 1983.

21 'Sapekoe women farm workers', *Agenda* no. 1 (Durban: Agenda Collective, 1987).

22 Ibid.

23 Reproduced with permission from *Schattendasein: Farmarbeiter in Sudafrika (Living in the Shadow — Farmworkers in South Africa)*, op. cit.

24 This section draws extensively on 'Organising farmworkers' by Andrew Ball, in *South African Labour Bulletin*, Vol. 14, No. 8, May 1990.

25 This section draws on an article entitled 'The land' by Jo-Anne Collinge, in *Work in Progress* 64 (Johannesburg, January 1990).

26 This quotation and the one following are taken with permission from *Umhlaba: Rural Land Struggles in the Transvaal in the '80s*, op. cit.

APPENDIX: Oxfam-supported organisations working with rural communities in South Africa

National projects

RSA 207: National Land Committee (previously National Committee Against Removals)

The NLC is the national co-ordinating body of five regional affiliates, engaged in support for urban and rural communities. NLC's main work is to publicise the plight of communities threatened with eviction and forced removal, and to campaign on their behalf. Recently its scope has broadened to include wider land-related issues, and its name has changed to reflect this shift. Since 1985 Oxfam has contributed £21,475 to the information and education work of NCAR/NLC.

RSA 285: Environmental and Development Agency

EDA works in rural areas throughout South Africa, offering organisational and technical training and advice to community groups. EDA also facilitates networking between communities; it offers information and resources on rural issues to the public; and conducts research on farm workers. Since 1986 Oxfam has contributed £30,604 to the Agency's work with rural communities, its resources centre, and its information work on environmental issues.

RSA 538: Project for the Study of Violence

Oxfam's role in this university-based research project is to pay the salaries of field workers who gather information on violence against farm workers and domestic workers. The project also runs a trauma clinic, which offers counselling to the victims of violence. Oxam will contribute a total of £47,079 from 1990 to 1993.

RSA 543: Legal Resources Centre of South Africa

The LRC's major concerns are human rights violations, labour matters, and problems of land and housing. In the Transvaal the LRC has focused on rural cases, such as farm workers' labour tenancy contracts, dismissals and evictions of farm worker families, the right to social pensions, and assaults by farmer employers. In 1990 Oxfam contributed £23,122 towards the running of the LRC's monthly legal clinics in rural areas.

RSA 551: Paper, Printing, Wood and Allied Workers Union

PPWAWU is striving to improve the working conditions of forestry workers, who are often as vulnerable and isolated as workers on private farms. Oxfam has contributed £11,494 to the union's education programme.

Regional projects

RSA 148: Transvaal Rural Aid and Development Programme

TRADP was set up in 1985 to help rural communities in the homeland areas bordering the Transvaal, particularly Bophuthatswana, with research, legal aid, and publicity. During the first two years Oxfam contributed £10,898 towards TRADP's legal aid and publicity work.

RSA 310: Transvaal Rural Action Committee

TRAC is an affiliate of the NLC, working with rural communities in the Transvaal. Since 1986 Oxfam has contributed £6,797 towards the expenses of a field worker based in Amsterdam, whose main area of work is with farm labourers, negotiating settlements in cases of eviction, and liaising with other rural communities and the LRC (see below).

RSA 310/608: Eastern Transvaal Network

This network of rural community committees is the result of the rural outreach work of TRAC and the LRC in the area. By coordinating information and para-legal training, the Network will be able to help farm workers and rural communities in a wider area. Oxfam paid the expenses of the first meetings of community representatives in early 1990 (£584), and will continue to support the Network.

RSA 380: Surplus People Project

SPP, an affiliate of the NLC, has been working since 1985 on land and housing issues in the urban and rural communities of Western Cape and Namaqualand. Since 1986 Oxfam has contributed £10,324 towards SPP's rural programme in Namaqualand, campaigning against the forced removals of rural communities, and the eviction of pastoralists for a proposed national park.

RSA 434: Association for Rural Advancement

AFRA is the Natal-based affiliate of the NLC. AFRA works with rural communities threatened with eviction, harassed 'black spot' communities, and evicted farm workers, offering para-legal support and publicity. Oxfam has contributed £7,470 towards AFRA's information work.

RSA 529: Southwell Mission, Diocese of Grahamstown

The Southwell Mission School caters for the children of farm workers on surrounding farms in an isolated area outside Grahamstown, in the Eastern Cape. It is also the focus for evening adult literacy classes for the farm workers themselves. Oxfam has contributed £6,713 towards the building of a classroom and the provision of a vehicle.

RSA 547: Grahamstown Rural Committee

The GRC is an affiliate of the NLC, based in the Eastern Cape. It works mainly with rural communities in the Border region who have suffered forced removal or incorporation into the 'homelands' of Ciskei and Transkei. Oxfam has contributed £9,500 towards the GRC's information and publicity work.

RSA 552: Christ Church Mission Fund, Kei Road, Eastern Cape

This relief fund was set up to support over 100 evicted farm workers needing temporary shelter. Oxfam's initial contribution amounted to £1,685.

FURTHER READING

Front Line Africa: The Right to a Future, by Susanna Smith (Oxford: Oxfam, 1990)
Fruit of the Vine: The Human Cost of South African Wine (London: Catholic Institute of International Relations, 1989)
No Place to Rest: Forced Removals and the Law in South Africa, ed. Christina Murray and Catherine O'Regan (Oxford: Oxford University Press, 1990)
Uprooting Poverty: The South African Challenge, by F. Wilson and M. Ramphele (Cape Town and Johannesburg: David Philip, 1989)